Lecture Notes in Electrical Engineering

Volume 65

Gourab Sen Gupta ·
Subhas Chandra Mukhopadhyay

Embedded Microcontroller Interfacing

Designing Integrated Projects

Springer

Gourab Sen Gupta
School of Engineering and Advanced Technology (SEAT)
Massey University (Turitea Campus)
Palmerston North
New Zealand
E-mail: G.SenGupta@massey.ac.nz

Subhas Chandra Mukhopadhyay
School of Engineering and Advanced Technology (SEAT)
Massey University (Turitea Campus)
Palmerston North
New Zealand
E-mail: S.C.Mukhopadhyay@massey.ac.nz

ISBN 978-3-642-26384-2 ISBN 978-3-642-13636-8 (eBook)

DOI 10.1007/978-3-642-13636-8

Softcover reprint of the hardcover 1st edition 2010

Typeset & Coverdesign: Scientific Publishing Services Pvt. Ltd., Chennai, India.

Printed on acid-free paper

9 8 7 6 5 4 3 2 1

springer.com

Preface

Mixed-Signal Embedded Microcontrollers are commonly used in integrating analog components needed to control non-digital electronic systems. They are used in automatically controlled devices and products, such as automobile engine control systems, wireless remote controllers, office machines, home appliances, power tools, and toys. Microcontrollers make it economical to digitally control even more devices and processes by reducing the size and cost, compared to a design that uses a separate microprocessor, memory, and input/output devices. In many undergraduate and post-graduate courses, teaching of mixed-signal microcontrollers and their use for project work has become compulsory.

Students face a lot of difficulties when they have to interface a microcontroller with the electronics they deal with. This book addresses some issues of interfacing the microcontrollers and describes some project implementations with the Silicon Lab C8051F020 mixed–signal microcontroller. The intended readers are college and university students specializing in electronics, computer systems engineering, electrical and electronics engineering; researchers involved with electronics based system, practitioners, technicians and in general anybody interested in microcontrollers based projects.

The complete book is divided into ten chapters. It is our view that expertise in microcontrollers is achieved by using it in different applications. Most of the book is dedicated to describe a few project implementations. Six different successful projects have been detailed.

Chapter 1 describes the fundamentals of electronics and analog processing circuits. The input signal is almost always passed through some analog processing circuits before it is interfaced to a microcontroller. For signal processing, the basic knowledge of this chapter is very important.

Chapter 2 gives an overview of the SiLab C8051F020 micro-controller. On-chip peripherals such as ADC and DAC, and other features like the digital crossbar and the voltage reference generator are briefly introduced. While programming using a high level language, such as C, makes it less important to know the intricacies of the hardware architecture of the microcontroller, it is still beneficial to have some knowledge of the memory organization and the special function registers.

Chapter 3 introduces the Keil™ C compiler for the SiLab C8051F020 microcontroller. The high level language C, in combination with some standard codes, is used to develop the software program. The differences in programming the C8051F020 in C, compared to a standard C program, are almost all related to architectural issues which are highlighted in this chapter.

Chapter 4 describes some of the important design issues of interfacing a microcontroller to common electronic circuits. Open collector configuration, loading problems, microcontroller cross-bar definition, driving output load etc. have been discussed.

In chapter 5 we have detailed the development of a DC motor control project using Silabs 8051F020. It has been taken directly off our teaching program. A problem based learning and teaching approach was taken and our main focus in this chapter is to describe in detail the laboratory exercise in which the students work in groups for a complete semester. The software part of the project involves the development of the software to control the speed of the DC motor. The hardware part of the project is the design and development of over-current protection circuit and the associated expansion board.

In chapter 6 we have described the design, fabrication and implementation of a switched mode power supply based on both discrete circuit and embedded microcontroller. Even though an integrated circuit (IC) with a complete switched mode power supply is now available, from a student's learning perspective still a lot of things can be learnt while doing this project.

Chapter 7 details the implementation of an embedded microcontroller based control system for magnetic levitation.

Chapter 8 describes the hardware implementation of a microcontroller based remote firing module to detonate fireworks and the software to control this remote module. Traditional systems for fireworks detonation usually use very long runs of cable, up to several hundred meters, for each firework connected. This increases the setup time and cost significantly. To reduce the amount of wiring, short lengths of cable are often used; this however places the technicians at risk because of the close proximity to the firework shells. The proposed wireless system overcomes these shortcomings.

Chapter 9 describes an embedded microcontroller based sensing system for seafood inspection. Interdigital sensors have been used for non-destructive and noninvasive inspection of the material properties. There are many applications of interdigital sensors based systems.

We are indebted to many of our students and colleagues who were involved with the various projects over several years and some of their works have been used in this book. In particular we would like to acknowledge the contribution of our past and present students Dan Paolo Salvador, Elijah Sheppard, Jamses Tingsley, Chinthaka Gooneratne, Anuroop Gaddam, Adam Bullen, Mohd. Syaifudin Abdul Rahman, Vishnu Kasturi, Karan Singh Malhi, Michelle Cho and Matthew Finnie. Chapters 2 and 3 are, in parts, reproduced by kind permission from Silicon Laboratories, USA, from the book "Embedded Programming with Field-Programmable Mixed-Signal Microcontrollers", Second Edition, 2008 (ISBN: 978-0-9800541-0-1) and we would like to thank Chew Moi Tin and Prof. Chris Messom who had contributed to it. Over the years we have received invaluable technical support for our projects from Ken Mercer and we thank him profusely.

We would also like to express our sincere thanks to our family members for their continuous support and patience.

We hope you find this book useful.

G. Sen Gupta
S. C. Mukhopadhyay
School of Engineering and Advanced Technology
Massey University (Manawatu)
Palmerston North, New Zealand

Contents

1

Operational Amplifier and Analog Signal Processing Circuits: A Revision

1.1 Introduction

In most embedded microcontroller based systems, the input signals of the microcontroller usually take a variety of forms in terms of magnitude, type, frequency and so on. On many occasions, it is not possible or advisable to connect the signals coming from sensors directly to the microcontroller. In some cases, even though the signal is in digital form, there is a need to change the level of voltage. In many situations the signals are in analog form and they need to go through a signal processing stage. In this chapter we will review the fundamental of electronic circuits which are mainly used as the signal processing circuits to interface signals from different transducers to microcontrollers.

An operational amplifier is a commonly used building block for implementing different signal processing circuits in the analog domain. In analog electronic circuits the operational amplifier is the most versatile device. The common IC amplifier is made up of a number of transistor stages on a single chip and is basically a voltage controlled voltage source. It is used as a fundamental building block in basic amplification, signal conditioning, active filters, function generators, switching capacitors etc. In its simplest form an operational amplifier, in short op-amp, is a three-terminal device with two inputs, inverting input V_- and non-inverting input V_+, and an output, V_{out} as shown in figure 1.1.

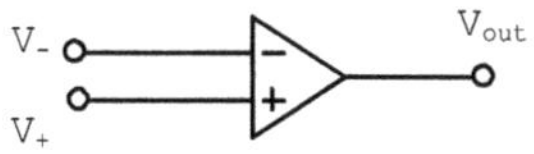

Fig. 1.1 Operational amplifier

Proper power supply should be provided to op-amp for its reliable operation. Operational amplifiers have the following properties-

1. an inverting input and a non-inverting input
2. very high (usually assumed infinite) input impedance at each input

G. Sen Gupta and S.C. Mukhopadhyay: Embedded Microcontroller Interfacing, LNEE 65, pp. 1–19.
springerlink.com

3. low output impedance
4. very large voltage gain (typically 10^5) when used in open-loop configuration (i.e. without feedback)
5. broad frequency bandwidth
6. free of drift due to change in ambient temperature
7. high stability.

Ideally the gain and the input impedance of the op-amp are infinite. In practical op-amps the input impedance is normally 100 MΩ or more. The output voltage is a function of the difference between the voltages at the input terminals.

$V_{out} = A(V_+ - V_-)$; where A is the open loop gain of the op-amp. For an ideal op-amp A is infinite.

From the above equation, we have, $(V_+ - V_-) = \frac{V_{out}}{A} \approx 0$; as A is infinite.

This means $V_+ = V_-$, i.e., even if the terminals are connected by a resistance, the current drawn will be zero. In other words the input terminals are virtually short-circuited as there is no potential difference between the two inputs. Since there is no current flowing, it means as if the terminals are not connected to each other. So the input terminals are virtually open-circuited. So the behavior of an ideal op-amp is summarized as-

1. current drawn by the op-amp at the input terminals is zero
2. output voltage is whatever makes the input terminal voltages equal.

For practical op-amp we need two terminals for the power supply, usually bipolar ± 12V or ± 15 V, though the op-amp can operate with only one supply. In the following sections a few commonly used circuits are described.

1.2 Voltage Follower Circuit

Figure 1.2 shows the circuit configuration of a voltage follower. It is also known as a unity gain amplifier, buffer amplifier or an isolation amplifier. Its main application arises due to the fact that its input resistance is very high so that it draws negligible current from the source. Many transducers provide a very small signal and it is very weak, so this configuration is ideal for those applications.

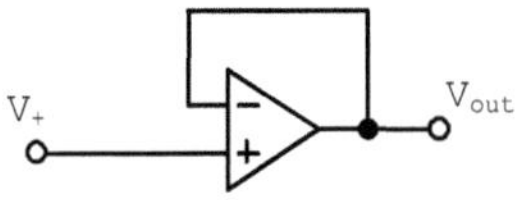

Fig. 1.2 Circuit configuration of a voltage follower

The input is fed directly to the +ve terminal. The output terminal and the inverting input terminal are shorted.

The relationship of the input and output voltage is $V_{out} = V_{+}$. The output voltage and the input voltage are equal in magnitude and have the same sign.

1.3 Inverting Amplifier

Figure 1.3 shows the circuit configuration of an inverting amplifier. The input is fed to the inverting input terminal through the resistance R_{in}. The non-inverting terminal is connected to ground. R_F is the feedback resistance and is connected between the output terminal and the inverting input terminal.

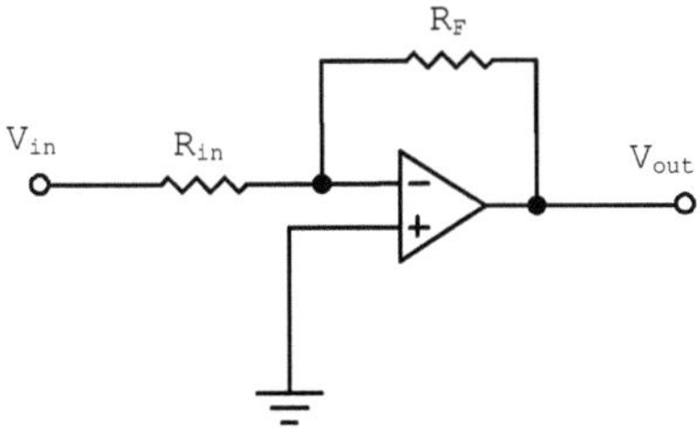

Fig. 1.3 Circuit configuration of an inverting amplifier

Assuming the current drawn at the inverting terminal is zero, the relationship between output and input is given by-

$$\frac{V_{in}}{R_{in}} + \frac{V_{out}}{R_F} = 0 \quad \text{which gives} \quad V_{out} = -\frac{R_F}{R_{in}} V_{in} = -A_{cl} V_{in}$$

where A_{cl} is the closed-loop gain of the amplifier.

So the gain is decided by the ratio of the feedback resistance to the input resistance. There is an inversion of the polarity i.e., the output voltage is the inverted version of the input voltage. In other words, there is a phase difference of 180° between the output and input voltages. By proper choice of R_F and R_{in} any value of output voltage can be obtained. The maximum value of the output voltage is, however, limited by the supply voltage of the amplifier.

1.4 Sign Changer

In figure 1.3 if $R_F = R_{in}$, then $V_{out} = -V_{in}$, i.e., the sign of the input has been changed without a change in magnitude. If two such amplifiers are connected in cascade, the sign of the input and the output are the same.

1.5 Phase Shifter

A phase shifter is obtained by substituting Z_F in place of R_F in figure 1.3. The input resistance may also be replaced by an impedance, Z_{in}. If the magnitudes of Z_F and Z_{in} are equal but the phase angles are different, the operational amplifier will shift the phase of the sinusoidal input signal without changing its magnitude. Any change in phase (0° to ± 180°) can be obtained.

1.6 Inverting Summing Amplifier

Figure 1.4 shows the circuit diagram for a summing amplifier.

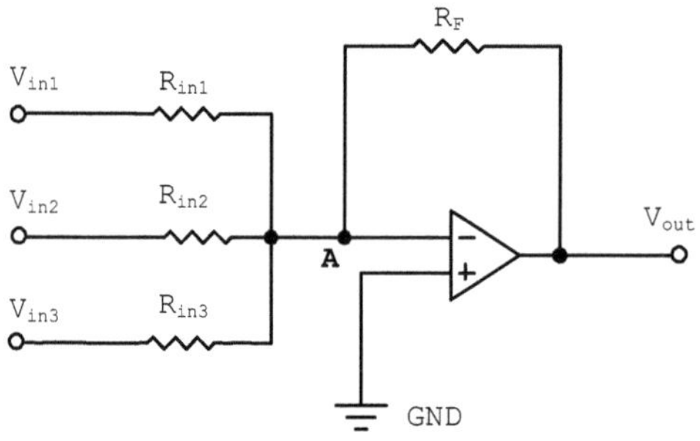

Fig. 1.4 Summing amplifier

Applying KCL at the node A and ignoring the input current to op-amp, we have

$$\frac{V_{in1}}{R_{in1}}+\frac{V_{in2}}{R_{in2}}+\frac{V_{in3}}{R_{in3}}+\frac{V_{out}}{R_F}=0$$

$$\text{Or,}V_{out}=-(\frac{R_F}{R_{in1}}V_{in1}+\frac{R_F}{R_{in2}}V_{in2}+\frac{R_F}{R_{in3}}V_{in3})$$

If $R_{in1}=R_{in2}=R_{in3}=R_F$, we get $V_{out}=-\left(V_{in1}+V_{in2}+V_{in3}\right)$

The output voltage is the sum of the input signals with phase reversed.

1.7 Non-inverting Amplifier

Figure 1.5 shows the circuit configuration of a non-inverting amplifier. The input impedance of this circuit is very high, ideally infinite.

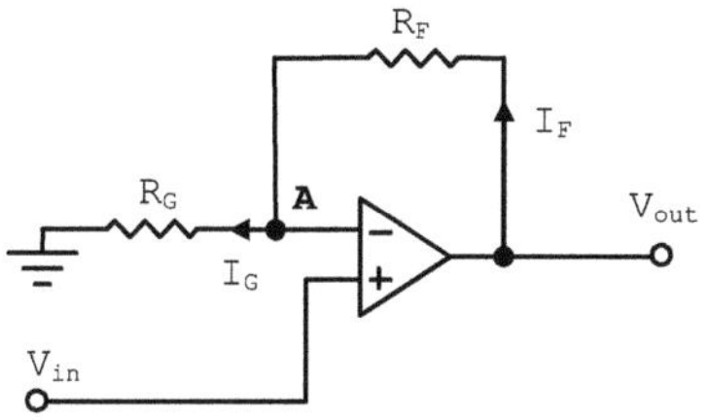

Fig. 1.5 Circuit configuration of non-inverting amplifier

Ideally, the voltage difference between the two terminals of the op-amp is zero. So $V_A = V_{in}$.

Applying KCL at node A, we have $I_F = I_G$

$$\frac{V_{out} - V_A}{R_F} = \frac{V_A}{R_G} \qquad \text{Or, } \frac{V_{out} - V_{in}}{R_F} = \frac{V_{in}}{R_G}$$

$$\text{or, } \frac{V_{out}}{R_F} = \frac{V_{in}}{R_F} + \frac{V_{in}}{R_G}$$

$$\text{or, } V_{out} = (\frac{R_F}{R_G} + 1)V_{in}$$

So, the voltage gain is given by, $A_{cl} = 1 + \frac{R_F}{R_G}$.

The current through the feedback resistance is given by,

$$I_F = \frac{V_{out} - V_A}{R_F} = \frac{V_{out} - V_{in}}{R_F} = \frac{V_{in}}{R_G}$$

It is seen that the current through R_F is independent of R_F and depends on V_{in} and R_G. This circuit can be used as a constant current source provided V_{in} and R_G remain constant with R_F the variable load. Also the circuit can be used as voltage controlled current source in which the current is proportional to the input voltage.

1.8 Non-inverting Summing Amplifier

In order to combine more than one input in non-inverting configuration the circuit, as shown in figure 1.6, can be used.

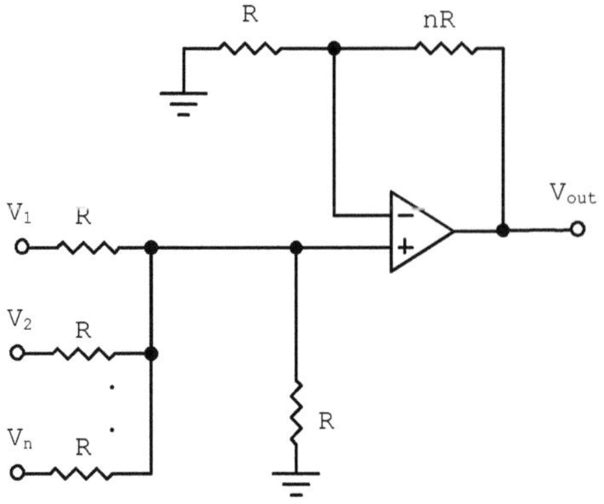

Fig. 1.6 Circuit configuration of non-inverting summing amplifier

The output voltage in terms of input voltages is given by,

$$V_{out} = V_1 + V_2 + V_3 + \ldots V_n.$$

1.9 Difference Amplifier

The output of a difference amplifier is the amplified version of the difference between two input signals. Figure 1.7 shows the circuit configuration. Any identical signal, for example noise, common to both the inputs is eliminated. The voltage at the input terminals A and B are given by,

$$v_A = V_{in1} - R_2 \frac{V_{in1} - V_{out}}{R_2 + R_1} \quad \text{and} \quad v_B = V_{in2} \frac{R_1}{R_1 + R_2}$$

Equating $v_A = v_B$, we have,

$$V_{in1} - V_{in1} \frac{R_2}{R_2 + R_1} + V_{out} \frac{R_2}{R_2 + R_1} = V_{in2} \frac{R_1}{R_2 + R_1}$$

$$Or, \quad \mathrm{V}_{out} = \frac{R_1}{R_2}(V_{in2} - V_{in1})$$

If $R_1 = R_2$, $V_{out} = V_{in2} - V_{in1}$, i.e., the difference between the two input signals.

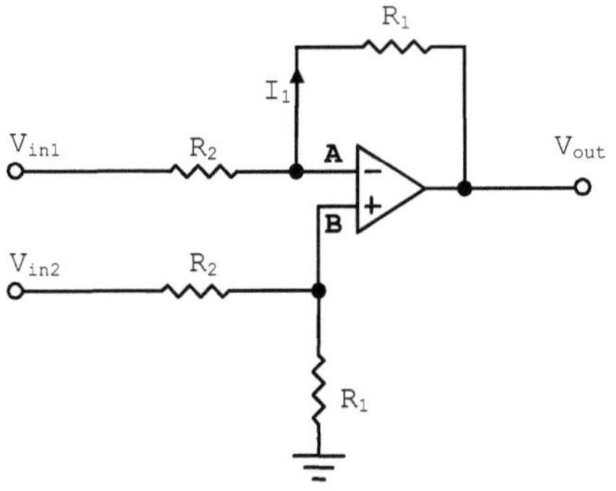

Fig. 1.7 Difference amplifier

1.10 Current to Voltage (I-V) Converter

Figure 1.8 shows the circuit configuration of a current to voltage converter.

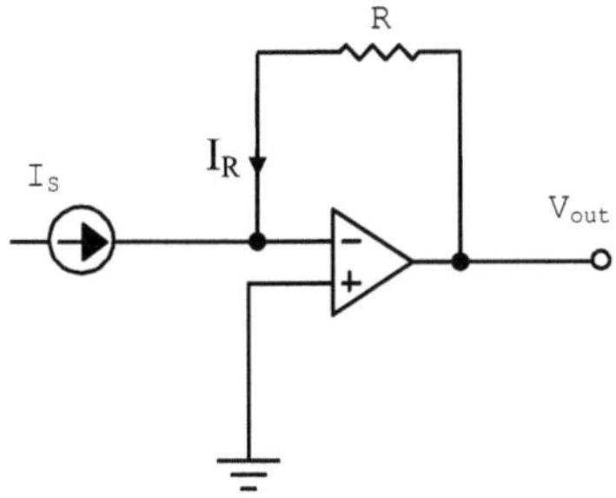

Fig. 1.8 Current to voltage converter

The current entering the inverting terminal is neglected. So the whole current is flowing through the resistor R.

$$I_S + I_R = 0 \quad \text{so, } I_S = -I_R$$

$$\text{So, } I_R = \frac{V_{out} - V_-}{R} \approx \frac{V_{out}}{R} \quad \text{or,} \quad I_S = -I_R = -\frac{V_{out}}{R} \quad \text{or,} \quad V_o = -I_S R$$

The output voltage is directly proportional to the input current.

1.11 Integrator

Figure 1.9 shows the circuit configuration which acts as an integrator.

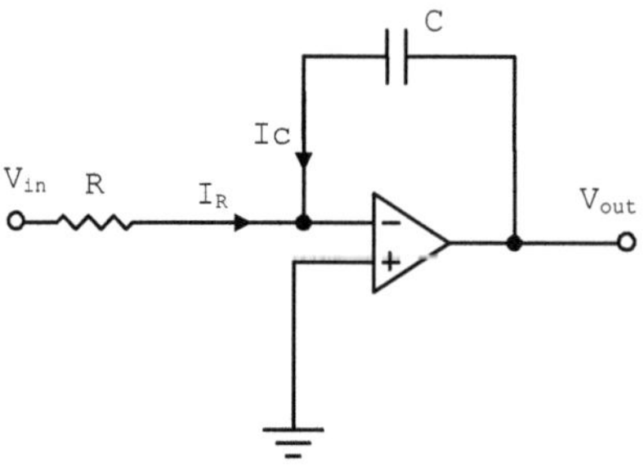

Fig. 1.9 An integrator

Applying KCL at inverting input terminal, we have $I_R + I_C = 0$

$$\frac{V_{in}}{R} + C\frac{d}{dt}(V_{out} - V_{\_}) = 0 \quad \text{Or,} \quad \frac{V_{in}}{R} + C\frac{d}{dt}V_{out} = 0$$

$$\frac{dV_{out}}{dt} = -\frac{V_{in}}{RC}$$

$$V_{out} = -\frac{1}{RC}\int_0^t V_{in}dt + V_{\text{intial}}$$

RC is the time constant of the integrator. For a discharged capacitor V_{initial} is zero. If any DC component exists in the input, the output will continuously increase which eventually results in the saturation of the output at one supply voltage. Sometimes to overcome this problem the circuit is usually modified to reduce its DC gain.

In some application the initial voltage across the capacitor may create problem. So the capacitor can be discharged by using a semiconductor switch as shown in figure 1.10. The capacitor gets discharged when the FET switch is closed. Constant input voltage V_{in} and regular pulses on Reset will produce a saw-tooth waveform.

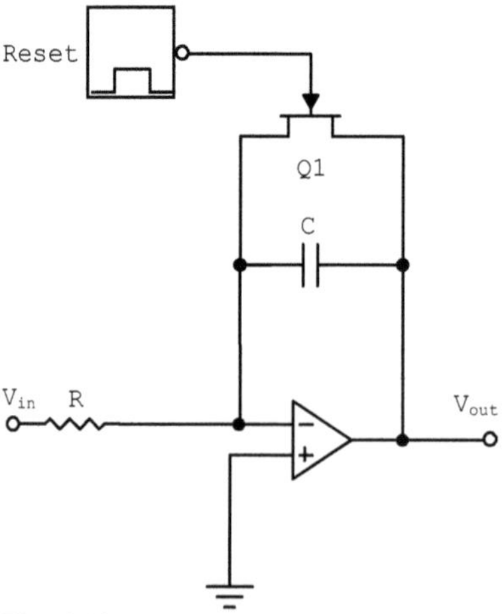

Fig. 1.10 An integrator with reset

1.12 Differentiator

By interchanging the relative position of the capacitor and the resistor in figure 1.9, a differentiator circuit is implemented. Figure 1.11 shows the circuit configuration of a differentiator.

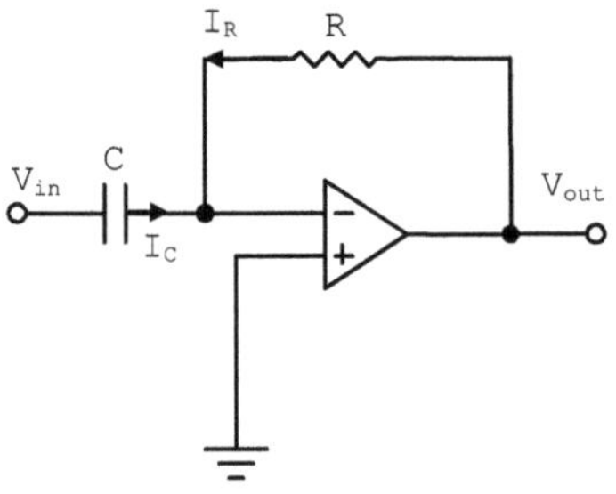

Fig. 1.11 A differentiator

Applying KCL at the inverting input terminal, we have $I_C + I_R = 0$

$$C\frac{d}{dt}(V_{in} - V_-) + \frac{V_{out} - V_-}{R} = 0$$

$$\text{or,} \quad V_{out} = -RC\frac{dV_{in}}{dt}$$

So the output voltage is proportional to the derivative of the input voltage with respect to time.

This circuit amplifies the high frequency signals and unwanted noise and therefore is inherently unstable. Connecting the non-inverting input to ground via a resistance R, reduces the effect of input bias current.

1.13 Comparators and Schmitt Triggers

In many applications, a special type of op-amp is used to compare a signal voltage on one of the input terminal with a reference voltage on the other input terminal, which is often adjustable.

Figure 1.12 shows the configuration of a comparator with the non-inverting (reference) input set to V_{REF}. The input is connected to the inverting input.

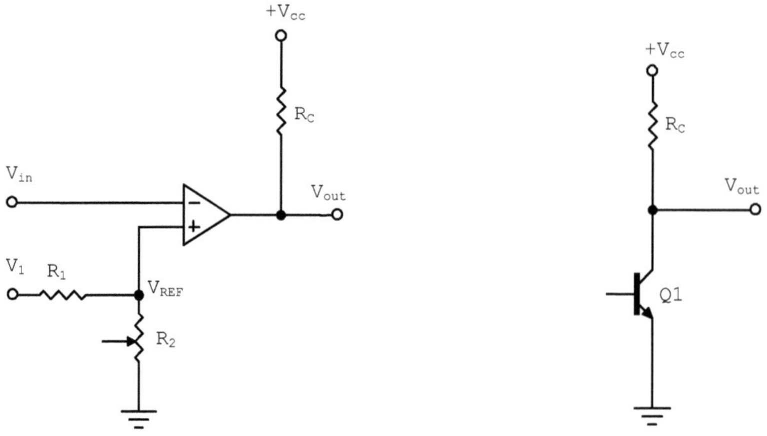

Fig. 1.12 Comparator configuration

It is called the open-collector configuration i.e., the collector of the transistor is kept open. The output of the comparator should be connected to the supply through a resistor, R_C, normally about 1 kΩ.

The reference voltage can be changed by changing the variable resistance R_2 and is given by-

$$V_{REF} = V_1 \frac{R_2}{R_1 + R_2}$$

When $V_{in} > V_{REF}$, $V_o = 0$ or $-V_{cc}$

$V_{in} < V_{REF}$, $V_o = V_{cc}$

R_C can be replaced by a relay, heater, lamp or motor.

When $V_{REF} = 0V$, the comparator becomes a zero-crossing detector.

Some IC comparators are-
LM 311
LM 339
LT 1016

Usually the comparators are fast acting and very often the input signals are noisy. There is a strong possibility that the output will undergo several on-off transitions as the input signal crosses the reference voltage level. This undesirable behavior can be prevented by using positive feedback from the output as shown in figure 1.13, which is called a Schmitt trigger. The output now has a hysteresis because the reference voltage, V_{REF}, depends on the output level.

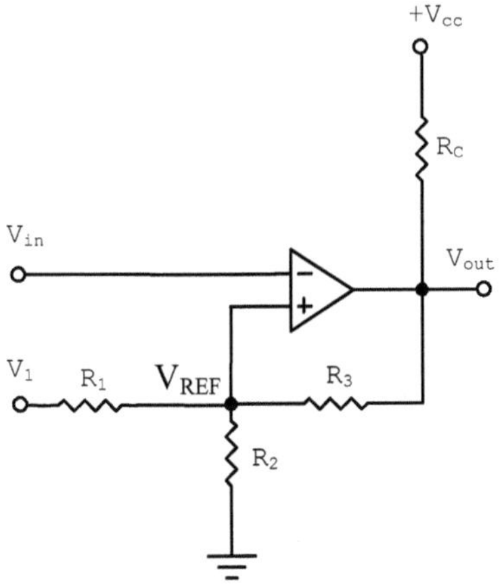

Fig. 1.13 Schmitt trigger

Applying KCL at the non-inverting input terminal i.e., the junction point of resistances R_1, R_2 and R_3

$$\frac{V_{REF}-V_1}{R_1}+\frac{V_{REF}}{R_2}+\frac{V_{REF}-V_{out}}{R_3}=0$$

$$\text{or,}\quad V_{REF}\left(\frac{1}{R_1}+\frac{1}{R_2}+\frac{1}{R_3}\right)=\frac{V_{out}}{R_3}+\frac{V_1}{R_1}$$

When V_{out} goes to +Vcc, V_{REF} goes to the higher V_{REF}
When V_o goes to –Vcc, V_{REF} goes to the lower V_{REF}.

So the lower and higher V_{REF} gives a reference band. By choosing appropriate resistance values the hysteresis band can be properly designed.

1.14 Logarithmic Amplifier

Figure 1.14 shows the circuit configuration for a logarithmic amplifier. A diode has been connected in the feedback loop.

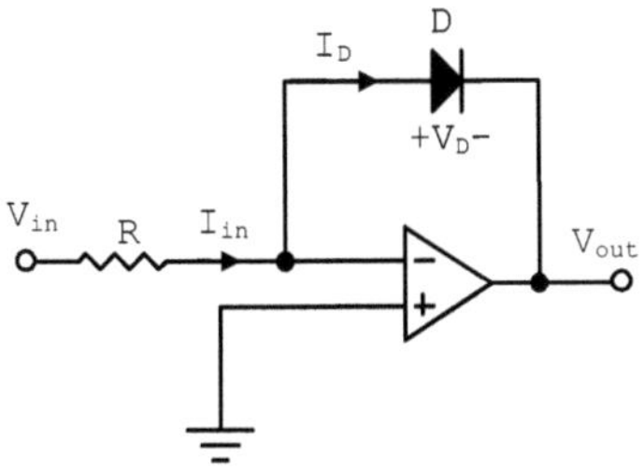

Fig. 1.14 Logarithmic amplifier

The relationship of the diode current and voltage is given by-

$$I_D = I_o(e^{V_D/\eta V_T} - 1) \approx I_o e^{V_D/\eta V_T}$$

where,

$V_T = \dfrac{kT}{e}$, and I_O is the reverse saturation current of the diode.

k is the Boltzmann's constant = $1.38e10^{-23}$ J/K
e is the charge of an electron = $1.6e10^{-19}$ C
T is the absolute temperature in Kelvin

At room temperature (300° K), the V_T, the voltage equivalent temperature of diode is:

$$V_T = \frac{1.38e10^{-23} * 300}{1.6e10^{-19}} = 26\text{mV}$$

Since the current drawn by the op-amp is negligible, we have-

$$I_{in} = \frac{V_{in}}{R} = I_D$$

So we can write,

$$V_{in} = RI_D = RI_o e^{V_D/\eta V_T}$$

Now, the output voltage $V_{out} = -V_D$ (the voltage drop across the diode).
Assuming $\eta = 1$ and replacing V_D by $-V_{out}$, we have,

$$V_{in} = RI_o e^{-V_{out}/V_T}$$

Taking logarithm of both sides,

$$\log v_{in} = \log(RI_o) - \frac{v_{out}}{V_T}$$

By properly choosing *R*, it is possible to make $RI_o = 1$, so the first term of the right hand side will be zero. So we can write,

$$\log v_{in} = -\frac{v_{out}}{V_T} \qquad \text{Or,} \quad v_{out} = -V_T \log v_{in}$$

1.15 Exponential Amplifier

Figure 1.15 shows the circuit configuration of an exponential (or antilog) amplifier. Since the current input to the op-amp is negligible, we have $I_D = I_R$. Because of the virtual ground at input, we have $V_{in} = V_D$.

$$I_R = I_D = I_o e^{V_D/\eta V_T} = I_o e^{Vin/\eta V_T}$$

Assuming $\eta = 1$ and $V_{out} = -I_R R$, we get $\qquad V_{out} = -I_D R = -RI_o e^{v_i/V_T}$

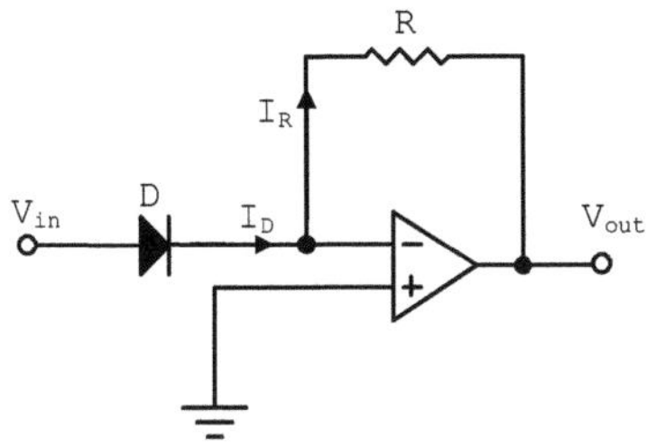

Fig. 1.15 Exponential amplifier

1.16 Single-Pole Filters

Figure 1.16 shows the circuit configuration of a low-pass filter. The capacitor C is represented by an impedance $1/(sC)$ in the s domain ($s = j\omega$).

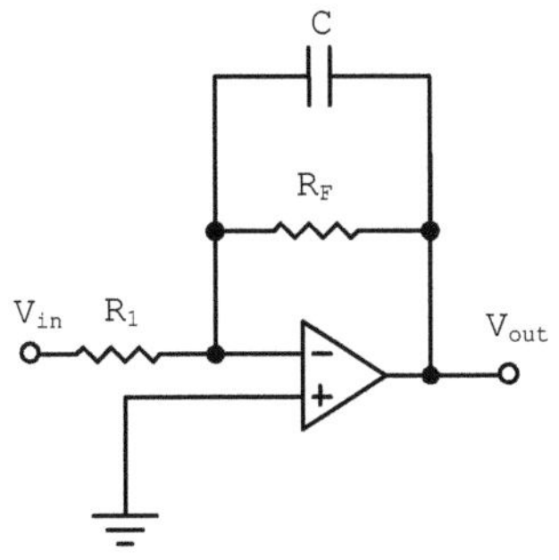

Fig. 1.16 Low pass filter with gain

$$Z_F(s) = \frac{R_F(1/sC)}{R_F + 1/sC} = \frac{R_F}{1 + sR_FC} \quad \text{and} \quad Z_1(s) = R_1$$

$$\text{So,} \quad V_o(s) = -\frac{Z_o(s)}{Z_1(s)} V_{in}(s) = -\frac{R_F / R_1}{1 + sR_FC} V_{in}(s)$$

where $V_o(s)$ and $V_{in}(s)$ are s-domain output and input signals.

The above equation represents a low-pass filter. When $sR_FC << 1$, the signals are passed with a gain of (R_F/R_1) without any attenuation. At $sR_FC = 1$, a 3 dB reduction in gain occurs. When $sR_FC >>1$, the signals are highly attenuated. The voltage gain decreases by 10 times (-20 dB) when the frequency increases by 10 times.

In many situations the voltage gain may not be required and a simple configuration is important. Figure 1.17 shows the configuration of the low pass filter with unity gain with a negative feedback loop. The configuration is a voltage follower and the filtering is achieved by the passive RC filter.

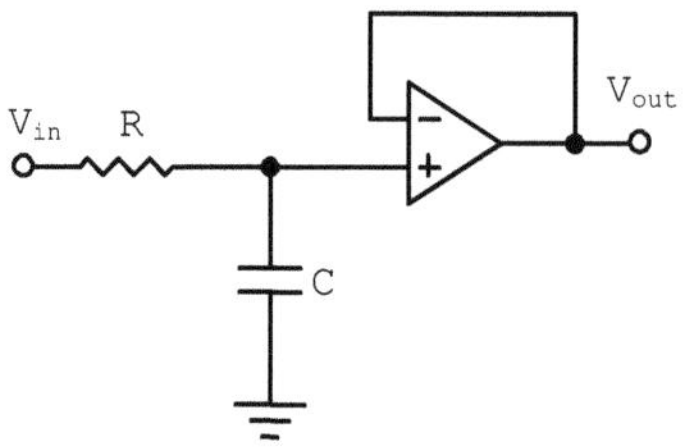

Fig. 1.17 Unity gain low pass filter

The expression of the output voltage is given by-

$$V_{out} = \frac{X_c}{\sqrt{R^2 + X_c^2}} V_{in}$$

The cut-off frequency of the configuration is given by-

$$\omega_c = \frac{1}{RC} \quad \text{or,} \quad f_c = \frac{1}{2\pi RC}$$

When the operating frequency $f << f_c$, the signals are passed through without any attenuation. At $f = f_c$, a 3 dB reduction in gain occurs. When $f >> f_c$, the signals are highly attenuated.

By interchanging the position of the resistor and the capacitor in figure 1.17, a high pass filter configuration is achieved. Figure 1.18 shows the configuration of a single-pole high pass filter based on passive RC filtering.

The expression of the output voltage is given by-

$$V_{out} = \frac{R}{\sqrt{R^2 + X_c^2}} V_{in}$$

The cut-off frequency of the configuration is given by-

$$\omega_c = \frac{1}{RC} \quad \text{Or,} \quad f_c = \frac{1}{2\pi RC}$$

When the operating frequency $f << f_c$, the signals are highly attenuated. At $f = f_c$, a 3 dB reduction in gain occurs. When $f >> f_c$, the signals are passed with a gain of unity without any attenuation.

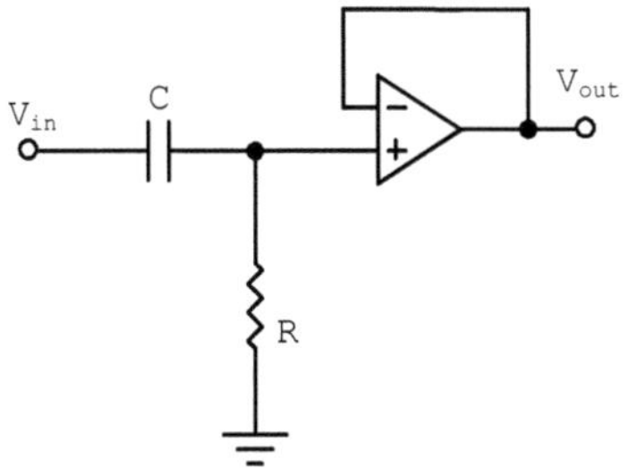

Fig. 1.18 Unity gain high pass filter

1.17 Double-Pole Filters

Figure 1.19 shows the configuration of a second order low pass filter.

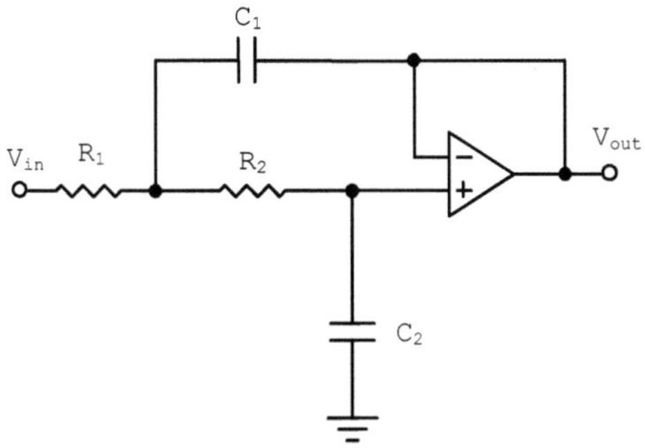

Fig. 1.19 Double-pole low pass filter

Figure 1.20 shows the configuration of a second order high pass filter.

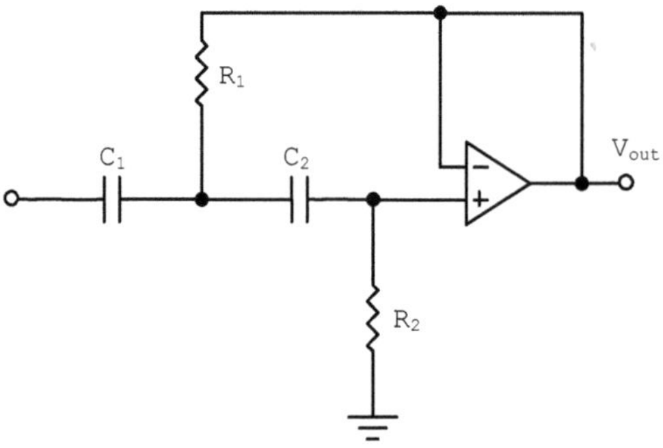

Fig. 1.20 Double-pole high pass filter

For both the circuits shown in figures 1.19 and 1.20, the critical frequency or the cut-off frequency is calculated by the following formula-

$$f_c = \frac{1}{2\pi\sqrt{R_1R_2C_1C_2}}$$

1.18 Band-Pass and Band-Stop Filters

By cascading a high pass filter with a low pass filter, it is possible to implement a band pass filter; the configuration for a double pole band pass filter is shown in figure 1.21.

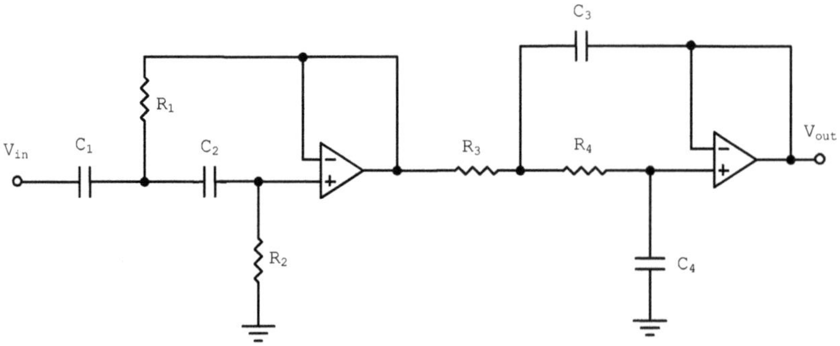

Fig. 1.21 Double-pole band pass filter

The critical frequency of each filter is chosen in such a way that their response curves overlap. The cut-off frequency of the high pass filter, f_{c1}, is lower than the cut-off frequency of the low pass filter, f_{c2}. The cut-off frequencies are given by-

$$f_{c1} = \frac{1}{2\pi\sqrt{R_1 R_2 C_1 C_2}} \qquad f_{c2} = \frac{1}{2\pi\sqrt{R_3 R_4 C_3 C_4}}$$

The centre frequency, f_c, of the pass band is the geometric average of f_{c1} and f_{c2}.

$$f_c = \sqrt{f_{c1} f_{c2}} \,.$$

It is also possible to implement a band-pass filter by using only one operational amplifier. Figure 1.22 shows the configuration of such as band-pass filter. The cut-off frequency of the filter is given by-

$$f_c = \frac{1}{2\pi CR}$$

The characteristics of the filter are affected by the gain of the filter. The gain is set by R_1 and R_2.

In many applications it may be necessary to stop signal of a particular band of frequencies. It is possible to implement a band-stop filter as shown in figure 1.23.

The cut-off frequency of the filter is given by-

$$f_c = \frac{1}{2\pi CR}$$

The characteristics of the filter are affected by the gain of the filter. The gain is set by R_1 and R_2.

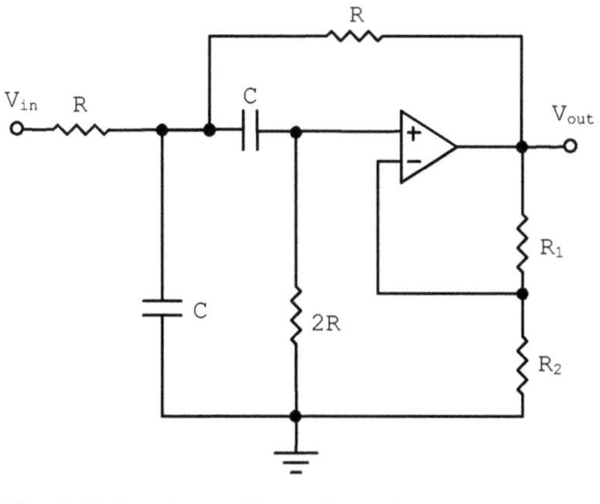

Fig. 1.22 Band-pass filter using only one op-amp

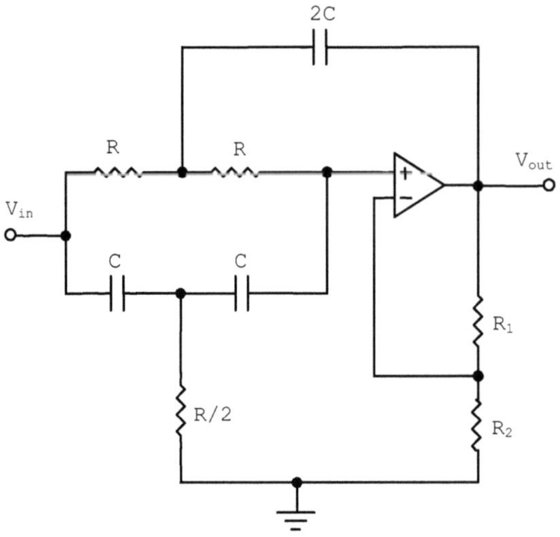

Fig. 1.23 Band-stop filter using only one op-amp

1.19 Oscillator Circuits

In many applications, oscillators are required to be implemented. Two op-amp based oscillators are described here. The first one provides sinusoidal output whereas the second one provides square wave output.

Figure 1.24 shows the configuration of a Wein-bridge oscillator. A series and parallel combination of resistors and capacitors are used for the feedback circuit. At the selected frequency of oscillation, the feedback circuit has a phase shift of zero and a gain of 1/3. So the network is configured as a non-inverting amplifier with a gain of 3. The gain of 3 is achieved by adjusting the feedback resistance; R_1 is two times the resistance R_2. If the gain is too low, the oscillation will stop. On the other hand a high gain will saturate the oscillator and the output will be distorted.

Figure 1.25 shows the circuit configuration of a simple op-amp based square wave generator. The circuit is also called as relaxation oscillator. The reference voltage at the non-inverting point A, depends on the output voltage and is given by-

$$V_{REF} = \pm V_{cc} \frac{R_1}{R_1 + R_2} = \pm V_{cc} \lambda$$

where λ is the feedback ratio. The supply voltage is ± V_{cc}.

The circuit is also called an astable multivibrator. The period of the square wave is given by-

$$T = 2RC \ln \frac{1+\lambda}{1-\lambda}$$

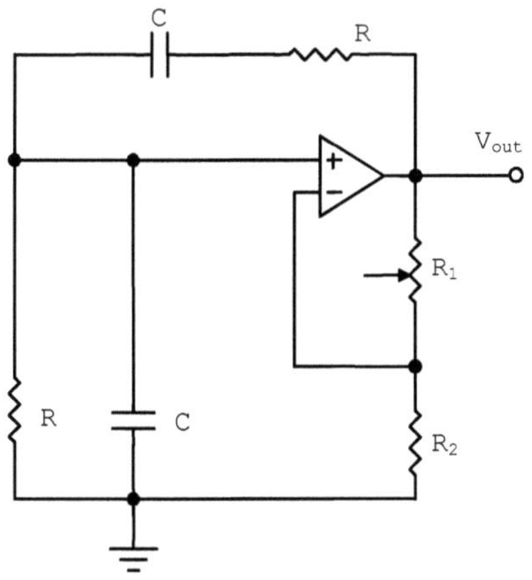

Fig. 1.24 Configuration of a Wein-bridge oscillator

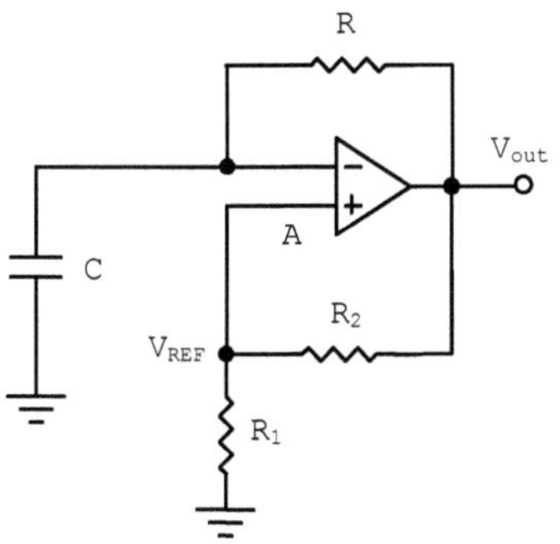

Fig. 1.25 Square wave generator

2

Introduction to Silicon Labs C8051F020 Microcontroller

2.1 Introduction

This chapter gives an overview of the SiLab C8051F020 micro-controller. On-chip peripherals like ADC and DAC, and other features like the crossbar and the voltage reference generator are briefly introduced. While programming using a high level language, such as C, makes it less important to know the intricacies of the hardware architecture of the micro-controller, it is still beneficial to have some knowledge of the memory organization and special function registers. Thus, these are also covered in this chapter.

2.2 CIP-51

SiLab mixed-signal system chips utilize the CIP-51 microcontroller core. The CIP-51 implements the standard 8051 organization, as well as additional custom peripherals. The block diagram of the CIP-51 is shown in figure 2.1.

The CIP-51 employs a pipelined architecture and is fully compatible with the MCS-51™ instruction set. The pipelined architecture greatly increases the instruction throughput over the 8051 architecture.

With the 8051, all instructions except for MUL and DIV take 12 or 24 system clock cycles to execute, and is usually limited to a maximum system clock of 12 MHz. By contrast, the CIP-51 core executes 70% of its instructions in one or two system clock cycles, with no instructions taking more than eight system clock cycles. With the CIP-51's maximum system clock at 25 MHz, it has a peak throughput of 25 millions of instructions per second (MIPS). The CIP-51 has a total of 109 instructions. Table 2.1 summarizes the number of instructions that require 1 to 8 clock cycles to execute.

G. Sen Gupta and S.C. Mukhopadhyay: Embedded Microcontroller Interfacing, LNEE 65, pp. 21–47.
springerlink.com

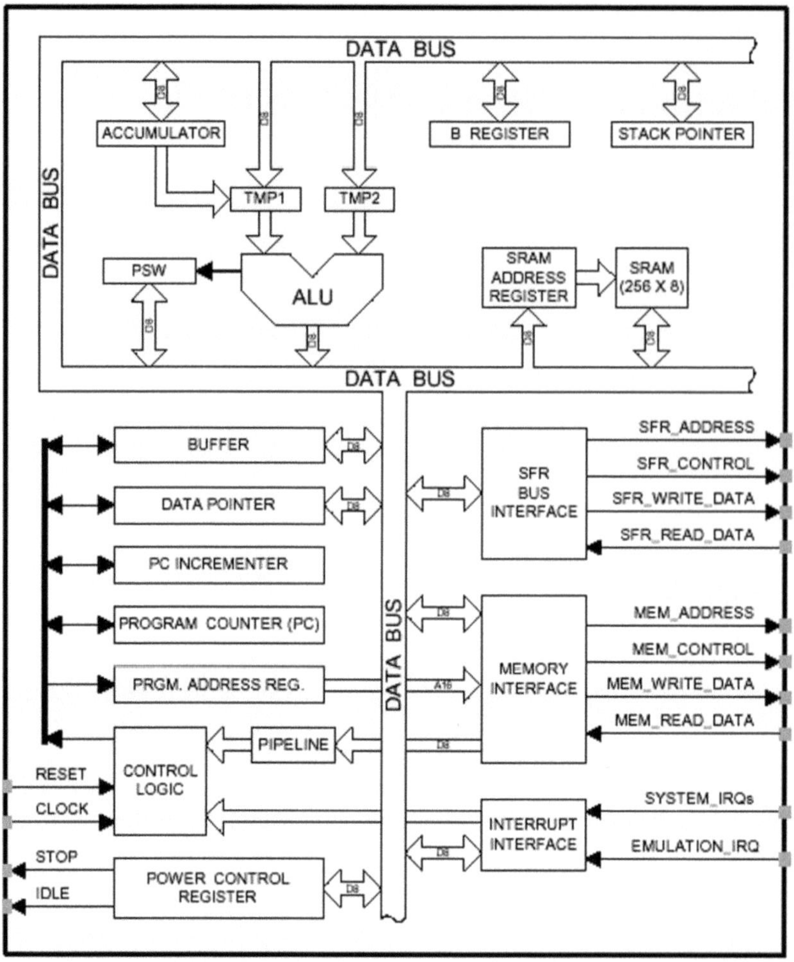

Fig. 2.1 Block diagram of CIP-51

Table 2.1 Execution Time of CIP-51 instructions

Clock Cycles to execute	1	2	2/3	3	3/4	4	4/5	5	8
Number of Instructions	26	50	5	14	7	3	1	2	1

2.3 C8051F020 System Overview

The SiLab C8051F020 is a fully integrated mixed-signal System-on-a-Chip microcontroller available in a 100 pin TQFP package. Its main features are shown in figure 2.2 and summarized in table 2.2. The block diagram is shown in figure 2.3.

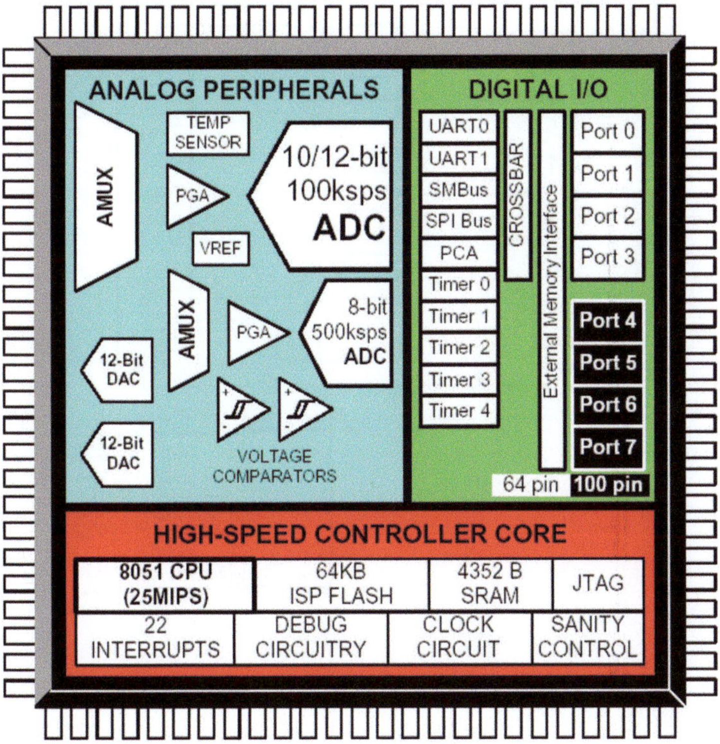

Fig. 2.2 System overview of the C8051F02x family

Table 2.2 C8051F020 Features

Peak Throughput	25 MIPS
FLASH Program Memory	64K
On-chip Data RAM	4352 bytes
Full-duplex UARTS	x 2
16-bit Timers	x 5
Digital I/O Ports	64 pins
12-bit 100ksps ADC	8 channels
8-bit 500ksps ADC	8 channels
DAC Resolution	12 bits
DAC Outputs	x 2
Analog Comparators	x 2
Interrupts	Two levels
Programmable Counter Arrays (PCA)	

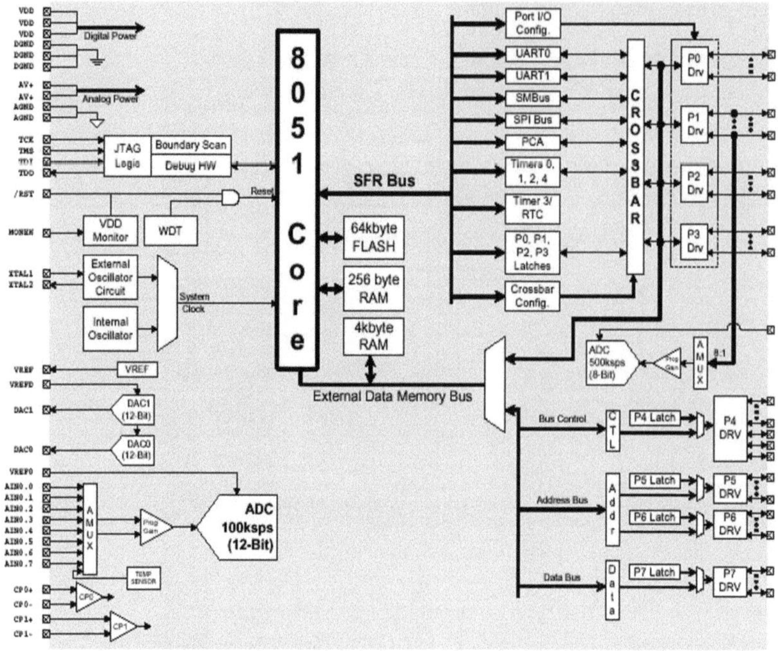

Fig. 2.3 Block diagram of C8051F020

All analog and digital peripherals are enabled/disabled and configured by user software. The FLASH memory can be reprogrammed even in-circuit, providing non-volatile data storage, and also allows field upgrades of the 8051 firmware.

2.4 Memory Organization

The memory organization of the CIP-51 System Controller is similar to that of a standard 8051 (Figure 1.2). There are two separate memory spaces: program memory and data memory. The CIP-51 memory organization is shown in figure 2.4. Program and data memory share the same address space but are accessed via different instruction types.

2.4.1 Program Memory

The C8051F020's program memory consists of 65536 bytes of FLASH, of which 512 bytes, from addresses 0xFE00 to 0xFFFF, are reserved for factory use. There is also a single 128 byte sector at address 0x10000 to 0x1007F (Scratchpad Memory), which is useful as a small table for software program constants.

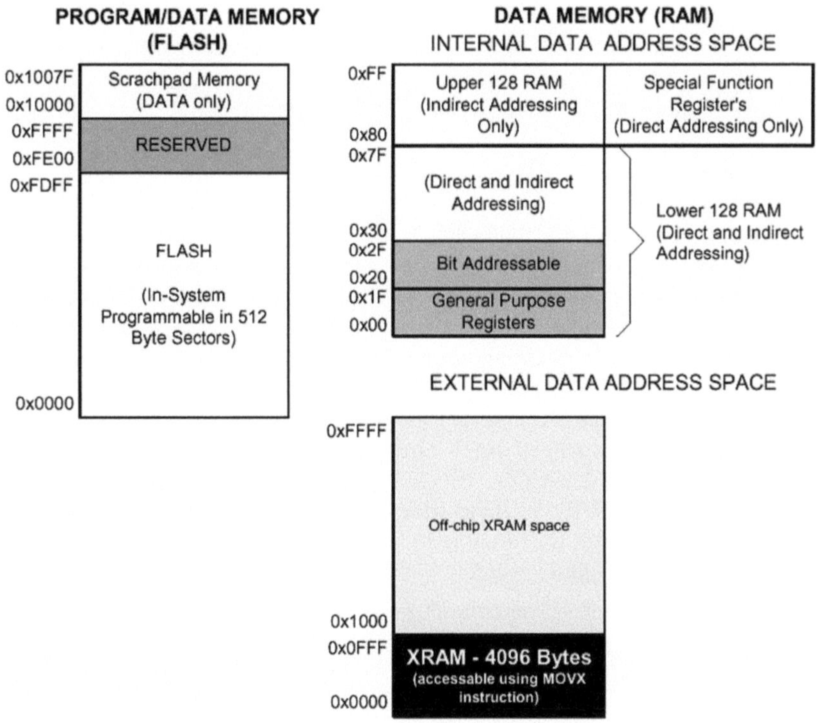

Fig. 2.4 C8051F020 memory map

2.4.2 *Data Memory*

The C8051F020 data memory has both internal and external address spaces. The internal data memory consists of 256 bytes of RAM. The Special Function Registers (SFR) are accessed anytime the direct addressing mode is used to access the upper 128 bytes of memory locations from 0x80 to 0xFF, while the general purpose RAM are accessed when indirect addressing is used (refer to Chapter 3 for addressing modes). The first 32 bytes of the internal data memory are addressable as four banks of 8 general purpose registers, and the next 16 bytes are bit-addressable or byte-addressable.

The external data memory has a 64K address space, with an on-chip 4K byte RAM block. An external memory interface (EMIF) is used to access the external data memory. The EMIF is configured by programming the EMI0CN and EMI0CF SFRs. The external data memory address space can be mapped to on-chip memory only, off-chip memory only, or a combination of the two (addresses up to 4K directed to on-chip, above 4K directed to EMIF). The EMIF is also capable of acting in multiplexed mode or non-multiplexed mode, depending on the state of the EMD2 (EMI0CF.4) bit.

2.4.3 *Stack*

The programmer stack can be located anywhere in the 256 byte internal data memory. A reset initializes the stack pointer (SP) to location 0x07; therefore, the first value pushed on the stack is placed at location 0x08, which is also the first register (R0) of register bank 1. Thus, if more than one register bank is to be used, the stack should be initialized to a location in the data memory not being used for data storage. The stack depth can extend up to 256 bytes.

2.4.4 *Special Function Registers (SFRs)*

The SFRs provide control and data exchange with the C8051F020's resources and peripherals. The C8051F020 duplicates the SFRs found in a typical 8051 implementation as well as implements additional SFRs which are used to configure and access the sub-systems unique to the microcontroller. This allows the addition of new functionalities while retaining compatibility with the MCS-51™ instruction set. Table 2.3 lists the SFRs implemented in the CIP-51 microcontroller.

The SFRs are accessed anytime the direct addressing mode is used to access memory locations from 0x80 to 0xFF. The SFRs with addresses ending in 0x0 or 0x8 (e.g. P0, TCON, P1, SCON, IE etc.) are bit-addressable as well as byte-addressable. All other SFRs are byte-addressable only. Unoccupied addresses in the SFR space are reserved for future use. Accessing these areas will have an indeterminate effect and should be avoided.

Table 2.3 SFR memory map

F8	SPI0CN	PCA0H	PCA0CPH0	PCA0CPH1	PCA0CPH2	PCA0CPH3	PCA0CPH4	WDTCN
F0	B	SCON1	SBUF1	SADDR1	TL4	TH4	EIP1	EIP2
E8	ADC0CN	PCA0L	PCA0CPL0	PCA0CPL1	PCA0CPL2	PCA0CPL3	PCA0CPL4	RSTSRC
E0	ACC	XBR0	XBR1	XBR2	RCAP4L	RCAP4H	EIE1	EIE2
D8	PCA0CN	PCA0MD	PCA0CPM0	PCA0CPM1	PCA0CPM2	PCA0CPM3	PCA0CPM4	
D0	PSW	REF0CN	DAC0L	DAC0H	DAC0CN	DAC1L	DAC1H	DAC1CN
C8	T2CON	T4CON	RCAP2L	RCAP2H	TL2	TH2		SMB0CR
C0	SMB0CN	SMB0STA	SMB0DAT	SMB0ADR	ADC0GTL	ADC0GTH	ADC0LTL	ADC0LTH
B8	IP	SADEN0	AMX0CF	AMX0SL	ADC0CF	P1MDIN	ADC0L	ADC0H
B0	P3	OSCXCN	OSCICN			P74OUT	FLSCL	FLACL
A8	IE	SADDR0	ADC1CN	ADC1CF	AMX1SL	P3IF	SADEN1	EMI0CN
A0	P2	EMI0TC		EMI0CF	P0MDOUT	P1MDOUT	P2MDOUT	P3MDOUT
98	SCON0	SBUF0	SPI0CFG	SPI0DAT	ADC1	SPI0CKR	CPT0CN	CPT1CN
90	P1	TMR3CN	TMR3RLL	TMR3RLH	TMR3L	TMR3H	P7	
88	TCON	TMOD	TL0	TL1	TH0	TH1	CKCON	PSCTL
80	P0	SP	DPL	DPH	P4	P5	P6	PCON
	0(8) Bit addressable	1(9)	2(A)	3(B)	4(C)	5(D)	6(E)	7(F)

2.5 I/O Ports and Crossbar

The standard 8051 Ports (0, 1, 2, and 3) are available on the C8051F020, as well as 4 additional ports (4, 5, 6, and 7) for a total of 64 general purpose port I/O pins. The port I/O behaves like the standard 8051 with a few enhancements. Access is possible through reading and writing the corresponding Port Data registers.

All port pins are 5 V tolerant, and support configurable Push-Pull or Open-Drain output modes and weak pull-ups. In addition, the pins on Port 1 can be used as Analog Inputs to ADC1. A block diagram of the port I/O cell is shown in figure 2.5.

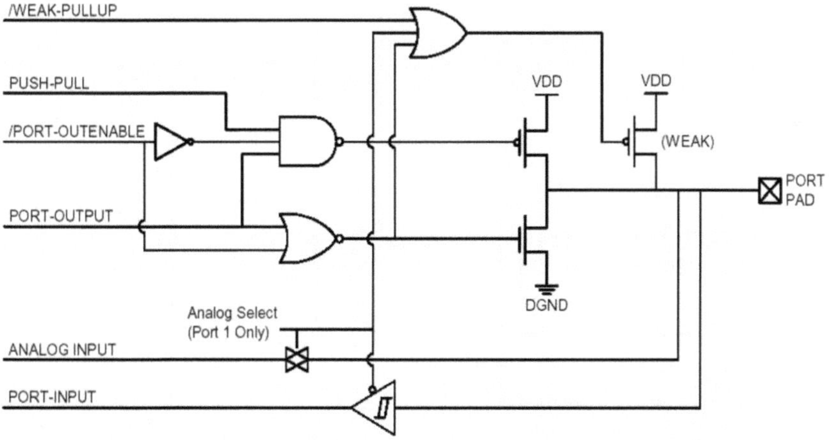

Fig. 2.5 Port I/O cell block diagram

The four lower ports (P0-P3) can be used as General-Purpose I/O (GPIO) pins or be assigned as inputs/outputs for the digital peripherals by programming a Digital Crossbar (Figure 2.6). The lower ports are both bit- and byte-addressable. The four upper ports (P4-P7) serve as byte-addressable GPIO pins.

The Digital Crossbar is essentially a large digital switching network that allows mapping of internal digital peripherals to the pins on Ports 0 to 3. The on-chip counter/timers, serial buses, HW interrupts, ADC Start of Conversion input, comparator outputs, and other digital signals in the controller can be configured to appear on the I/O pins by configuring the Crossbar Control registers XBR0, XBR1 and XBR2. This allows the system designer to select the exact mix of GPIO and digital resources needed for the particular application, limited only by the number of pins available. Unlike microcontrollers with standard multiplexed digital I/O, all combinations of functions are supported.

The digital peripherals are assigned Port pins in a priority order, starting with P0.0 and continue through P3.7 if necessary. UART0 has the highest priority and CNVSTR has the lowest priority.

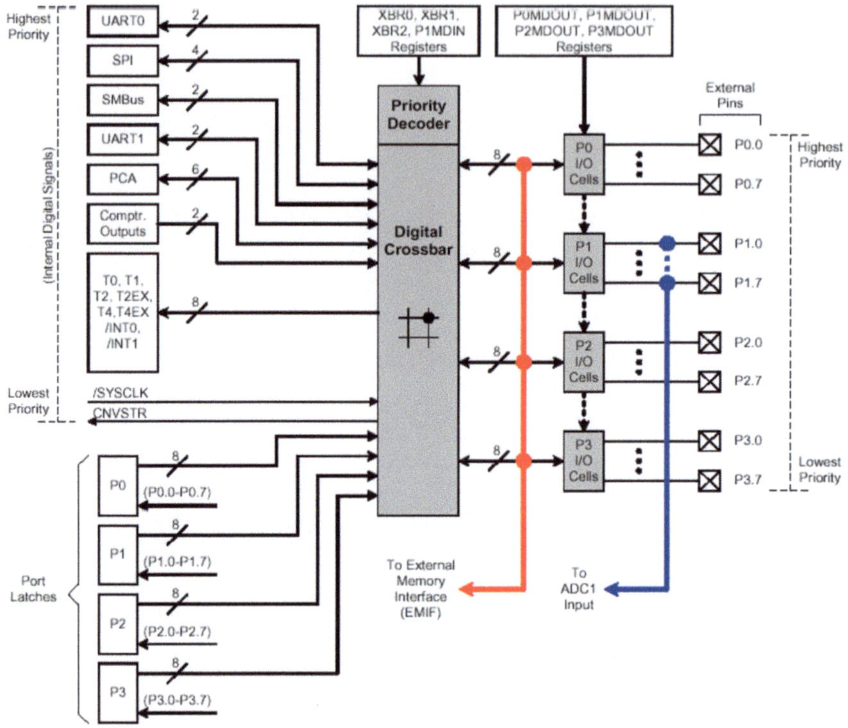

Fig. 2.6 Block diagram of lower port I/O (P0 to P3)

2.6 12-Bit Analog to Digital Converter

The C8051F020 has an on-chip 12-bit successive approximation register (SAR) Analog to Digital Converter (ADC0) with a 9-channel input multiplexer and programmable gain amplifier (Figure 2.7). A voltage reference is required for ADC0 to operate and is selected between the DAC0 output and an external VREF pin.

The ADC is configured via its associated Special Function Registers. One input channel is tied to an internal temperature sensor, while the other eight channels are available externally. Each pair of the eight external input channels can be setup as either two single-ended inputs or a single differential input.

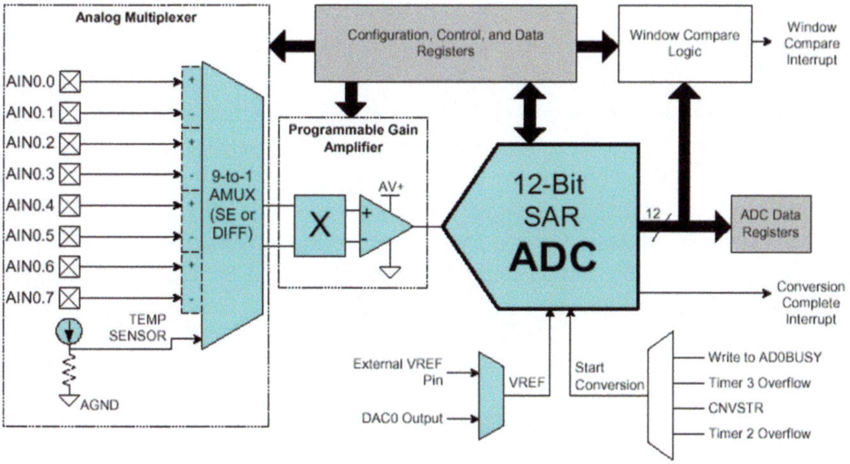

Fig. 2.7 12-Bit ADC block diagram

A programmable gain amplifier follows the analog multiplexer. The gain can be set in software from 0.5 to 16 in powers of 2. The gain stage is useful when different ADC input channels have widely varied input voltage signals, or when "zooming in" on a signal with a large DC offset (in differential mode, a DAC could be used to provide the DC offset).

Conversions can be started in four ways:

1. Software command,
2. Overflow of Timer 2,
3. Overflow of Timer 3, or
4. External signal input (CNVSTR).

Conversion completions are indicated by a status bit and an interrupt (if enabled). The resulting 12 bit data word is latched into two SFRs upon completion of a conversion. The data can be right or left justified in these registers (since ADC output is 12 bits but the two SFRs are 16 bits) under software control.

The Window Compare registers for the ADC data can be configured to interrupt the controller when ADC data is within or outside of a specified range. The ADC can monitor a key voltage continuously in background mode, but not interrupt the controller unless the converted data is within the specified window.

2.7 8-Bit Analog to Digital Converter

The C8051F020 has an on-board 8-bit SAR Analog to Digital Converter (ADC1) with an 8-channel input multiplexer and programmable gain amplifier (Figure 2.8).

Eight input pins are available for measurement. The ADC is again configurable via the SFRs. The ADC1 voltage reference is selected between the analog power supply (AV+) and an external VREF pin.

A programmable gain amplifier follows the analog multiplexer. The gain can be set in software to 0.5, 1, 2, or 4. Just as with ADC0, the conversion scheduling system allows ADC1 conversions to be initiated by software commands, timer overflows or an external input signal. ADC1 conversions may also be synchronized with ADC0 software-commanded conversions. Conversion completions are indicated by a status bit and an interrupt (if enabled), and the resulting 8 bit data word is latched into a SFR upon completion.

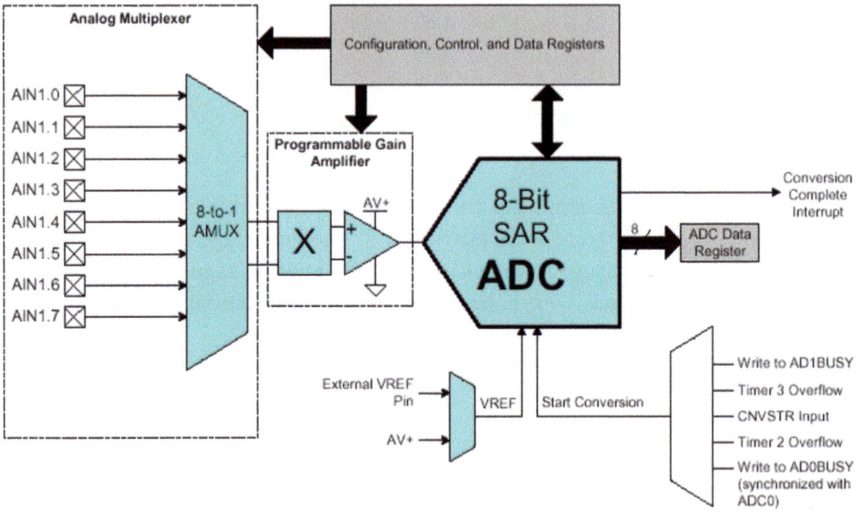

Fig. 2.8 8-Bit ADC block diagram

2.8 Digital to Analog Converters

The C8051F020 has two 12-bit Digital to Analog Converters, DAC0 and DAC1, as shown in figure 2.9. The DAC voltage reference is supplied via the dedicated VREFD input pin. The DAC output is updated each time when there is a software write (DACxH), or a Timer 2, 3, or 4 overflow (Figure 2.10). The DACs are especially useful as references for the comparators or offsets for the differential inputs of the ADC.

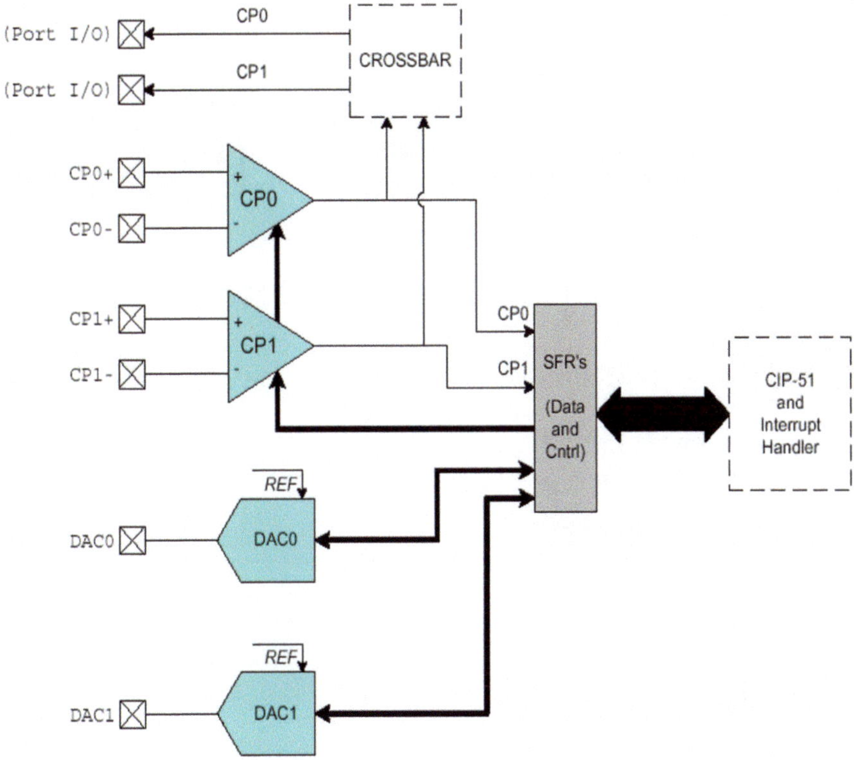

Fig. 2.9 Comparator and DAC block diagram

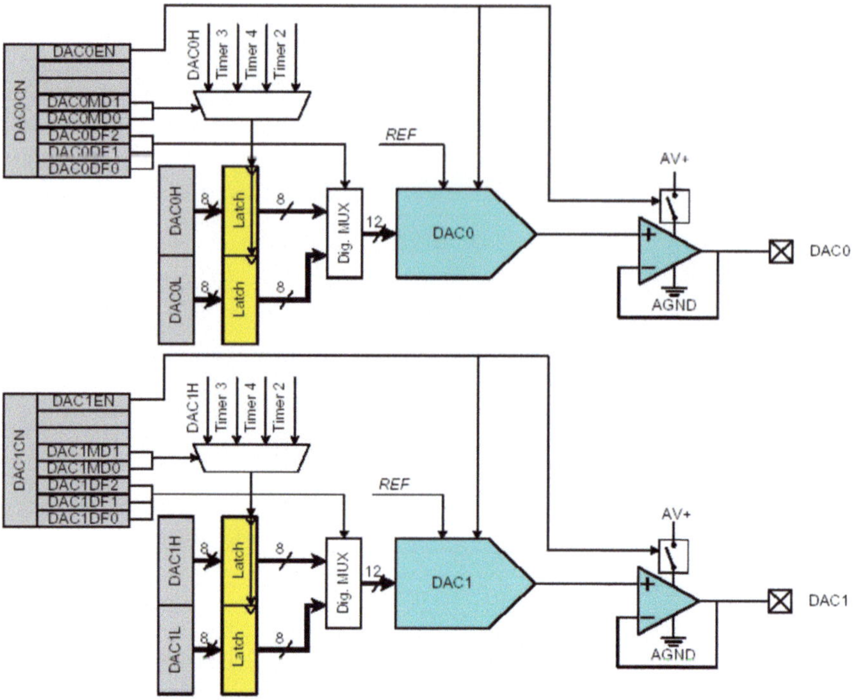

Fig. 2.10 DAC block diagram

2.9 Analog Voltage Comparators

There are two on-board voltage comparators, *Comparator0* and *Comparator1*, with software programmable hysteresis. The block diagram of the voltage comparators is shown in figure 2.11. The inputs of each comparator are available at the package pins. The outputs of the comparators are optionally available at the lower port I/O pins via the Crossbar. When assigned to package pins, each comparator output can be programmed to operate in open drain or push-pull modes. *Comparator0* can also be programmed as a reset source. The operation of *Comparator1* is identical to that of *Comparator0*, though *Comparator1* may not be configured as a reset source. The comparators can generate an interrupt on its rising edge, falling edge, or both. The comparators' output state can also be polled in software.

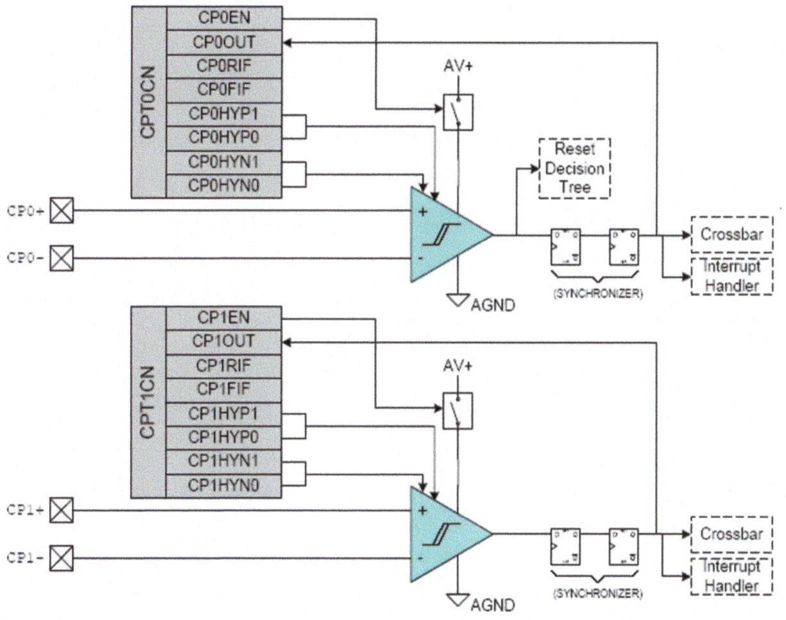

Fig. 2.11 Functional block diagram of the voltage comparators

2.9.1 Enable/Disable Comparator

There is one Comparator Control register (CPT0CN and CPT1CN for Comparator0 and Comparator1, respectively) for each comparator. Each comparator can be individually enabled or disabled (shutdown). Comparator0 is enabled by setting the CP0EN bit to logic 1, and is disabled by clearing this bit to logic 0.When disabled, the comparator output (if assigned to a Port I/O pin via the Crossbar) defaults to the logic low state, its interrupt capability is suspended and its supply current falls to less than 1 μA. Comparator inputs can be externally driven from -0.25 V to (AV+) +25 V without damage or upset.

2.9.2 Programmable Hysteresis

The hysteresis of each comparator is software-programmable. The user can program both the amount of hysteresis voltage (referred to the input voltage) and the positive and negative-going symmetry of this hysteresis around the threshold voltage. For the Comparator0, the hysteresis is programmed using bits 3-0 in the Comparator0 Control Register CPT0CN. The amount of negative hysteresis voltage is determined by the settings of the CP0HYN bits; in a similar way, the amount of positive hysteresis is determined by the setting the CP0HYP bits. The hysteresis plot is shown in figure 2.12. VIN- and VIN+ are the input signals at CP0-/CP1- and CP0+/CP1+ respectively.

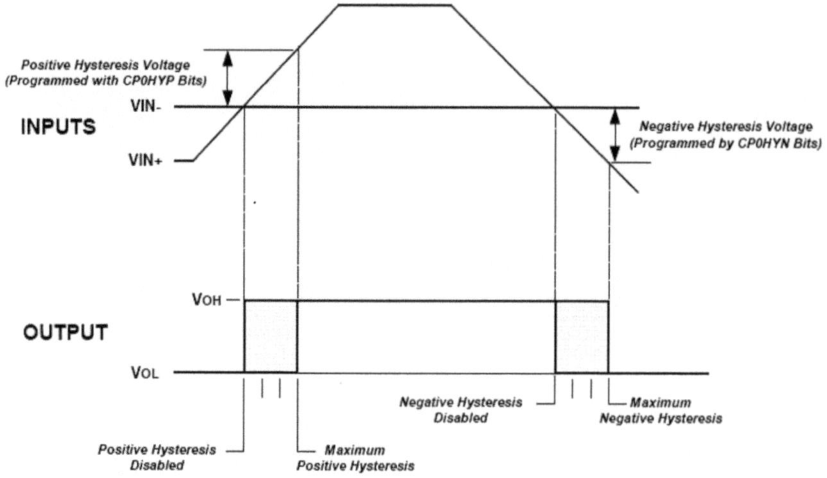

Fig. 2.12 Comparator hysteresis plot

2.9.3 Comparator Output and Interrupt

The output of the comparator can be polled in software, or can be used as an interrupt source. The output state of the Comparator can be obtained at any time by reading the CP0OUT (CP1OUT) bit. Comparator interrupts can be generated on rising-edge and/or falling-edge output transitions. The CP0FIF (CP1FIF) flag is set upon a Comparator falling-edge interrupt, and the CP0RIF (CP1RIF) flag is set upon the Comparator rising-edge interrupt. Once set, these bits remain set until cleared by software.

The Comparator0 falling edge interrupt and rising edge interrupts are enabled by setting ECP0F (EIE1.4) bit and ECP0R (EIE1.5) bit respectively. The priority for Comparator0 falling edge interrupt and rising edge interrupt are set by PCP0F (EIP1.4) bit and PCP0R (EIP1.5) bit respectively. Likewise, the Comparator1 falling edge interrupt and rising edge interrupts are enabled by setting ECP1F (EIE1.6) bit and ECP1R (EIE1.7) bit respectively. The priority for Comparator1 falling edge interrupt and rising edge interrupt are set by PCP1F (EIP1.6) bit and PCP1R (EIP1.7) bit respectively.

The following program shows how to set up the Comparator1 to generate interrupts on the falling and rising edges.

```
void Init_Comp1(void) //-- Initialise Comparator1
{
  //-- Enable Comparator1, 10mV positive and
  //  negative hysteresis
        CPT1CN = 0x8F;
  //-- Enable CP1 rising and falling edge interrupt
        EIE1 |= 0xC0;
}

//-- Comparator1 Rising Edge Interrupt Service Routine
void CP1RIF_ISR(void) interrupt 13
{
        //-- This interrupt is generated when CP1+ > CP1-
        CPT1CN &= 0xDF;          //-- clear the interrupt flag
        P5 = 0x10;               //-- turn on LED at P5.4
}

//-- Comparator1 Falling Edge Interrupt Service Routine
void CP1FIF_ISR(void) interrupt 12
{
        //-- This interrupt is generated when  CP1+ < CP1-
        CPT1CN &= 0xEF;          //-- clear the interrupt flag
        P5 = 0x20;               //-- turn on LED at P5.5
}
```

2.10 Voltage Reference

A voltage reference has to be used when operating the ADC and DAC. The C8051F020's three voltage reference input pins allow each ADC and the two DACs to reference an external voltage reference or the on-chip voltage reference output. ADC0 may also reference the DAC0 output internally, and ADC1 may reference the analog power supply voltage (AV+), via the VREF multiplexers shown in figure 2.13.

The internal voltage reference circuit consists of a 1.2V bandgap voltage reference generator and a gain-of-two output buffer amplifier, i.e. VREF is 2.4 V. The internal reference may be routed via the VREF pin to external system components or to the voltage reference input pins shown in figure 2.13. Bypass capacitors of 0.1 μF and 4.7 μF are recommended from the VREF pin to AGND.

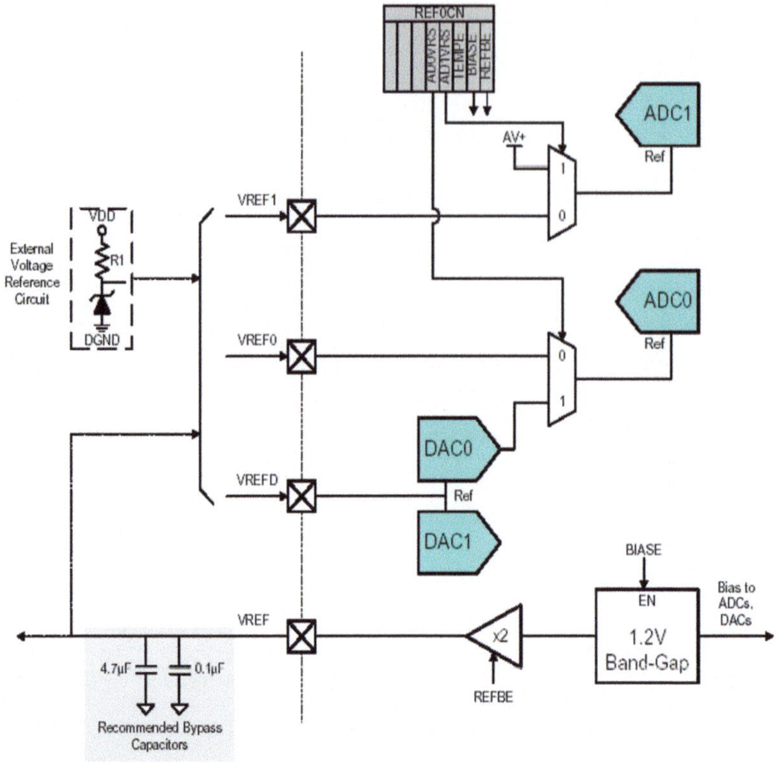

Fig. 2.13 Voltage reference functional block diagram

2.10.1 REF0CN: Reference Control Register

The Reference Control Register, REF0CN, enables/disables the internal reference generator and selects the reference inputs for ADC0 and ADC1 (Table 2.4).

Example: `MOV REF0CN, #00000011B`

This enables the use of the ADC or DAC, and the internal voltage reference.

Table 2.4 REF0CN: Reference Control Register

Bit	Symbol	Description
7-5	-	Unused. Read=000b; Write=Don't care.
4	AD0VRS	***ADC0 Voltage Reference Select*** 0: ADC0 voltage reference from VREF0 pin. 1: ADC0 voltage reference from DAC0 output.
3	AD1VRS	***ADC1 Voltage Reference Select*** 0: ADC1 voltage reference from VREF1 pin. 1: ADC1 voltage reference from AV+
2	TEMPE	***Temperature Sensor Enable Bit*** 0: Internal Temperature Sensor Off. 1: Internal Temperature Sensor On.
1	BIASE	***ADC/DAC Bias Generator Enable Bit*** **(Must be '1' if using ADC or DAC)** 0: Internal Bias Generator Off. 1: Internal Bias Generator On.
0	REFBE	***Internal Reference Buffer Enable Bit*** 0: Internal Reference Buffer Off. 1: Internal Reference Buffer On. Internal voltage reference is driven on the VREF pin.

2.11 Programmable Counter Array (PCA)

The Programmable Counter Array (PCA0) provides enhanced timer functionality while requiring less CPU time. PCA0 consists of a dedicated 16-bit counter/timer and five 16-bit capture/compare modules. Each capture/compare module has its own associated I/O line (CEXn) which is routed through the Crossbar to Port I/O when enabled. The counter/timer is driven by a programmable time-base that can select between six inputs as its source which are:

- system clock
- system clock divided by four
- system clock divided by twelve
- external oscillator clock source divided by 8
- Timer 0 overflow
- external clock signal on the ECI line

Each capture/compare module may be configured to operate independently in one of the following six modes:

- Edge-Triggered Capture
- Software Timer

- High-Speed Output
- Frequency Output
- 8-Bit PWM
- 16-Bit PWM

The PCA0 is configured and controlled through the system controller's Special Function Registers (SFRs). The basic block diagram of the PCA is shown in figure 2.14.

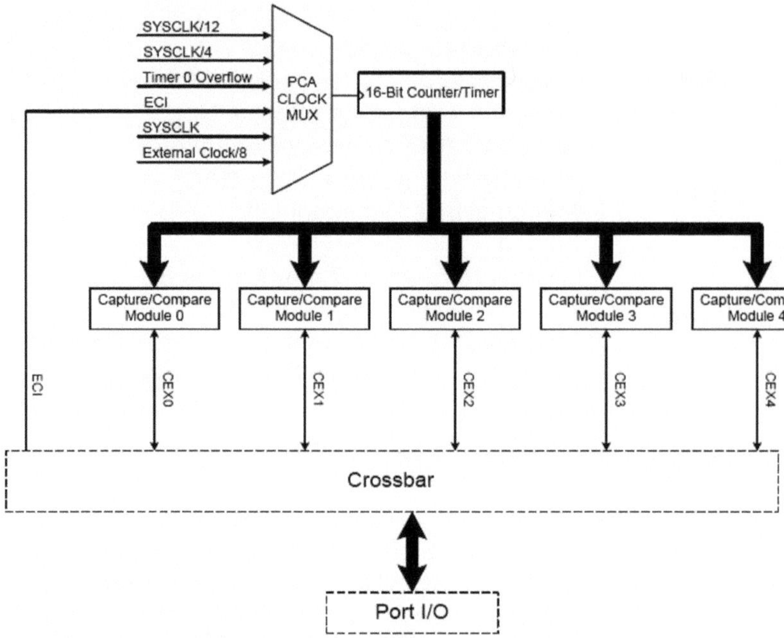

Fig. 2.14 Block diagram of the Programmable Counter Array

2.11.1 PCA Counter/Timer and Timebase Selection

The 16-bit PCA counter/timer consists of two 8-bit SFRs: PCA0L (low byte) and PCA0H (high byte). When reading the 16-bit counter value, the low byte must be read first, followed by the high byte to guarantee an accurate reading of the entire 16-bit PCA0 counter. Reading PCA0L automatically latches the value of PCA0H into a "snapshot" register; the following PCA0H read accesses this "snapshot" register. Reading PCA0H or PCA0L does not disturb the counter operation. The CPS2-CPS0 bits in the PCA0MD register select the timebase for the counter/timer as shown in table 2.5. Figure 2.15 shows the block diagram of the PCA Counter/Timer.

Table 2.5 PCA timebase selection

CPS2	CPS1	CPS0	Timebase
0	0	0	System clock divided by 12
0	0	1	System clock divided by 4
0	1	0	Timer 0 overflow
0	1	1	High-to-low transitions on ECI (max rate = system clock divided by 4)
1	0	0	System clock
1	0	1	External oscillator source divided by 8

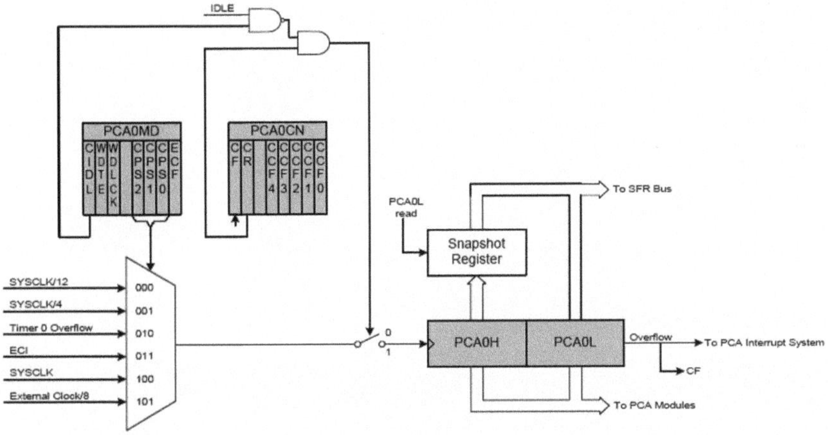

Fig. 2.15 Block diagram of the PCA Counter/Timer

When the counter/timer overflows from 0xFFFF to 0x0000, the Counter Overflow Flag (CF) in PCA0CN (PCA0 Control register) is set to logic 1 and an interrupt request is generated if CF interrupts are enabled. Setting the ECF bit in PCA0MD (PCA0 Mode register) to logic 1 enables the CF flag to generate an interrupt request. The CF bit is not automatically cleared by hardware when the CPU vectors to the interrupt service routine, and must be cleared by software. PCA0 interrupts must be globally enabled before CF interrupts are recognized. PCA0 interrupts are globally enabled by setting the EA bit (IE.7) and the EPCA0 bit (EIE1.3) to logic 1. Clearing the CIDL bit in the PCA0MD register allows the PCA to continue normal operation while the CPU is in *Idle* mode.

2.11.2 Operation Modes and Interrupts

As mentioned earlier, each capture/compare module can be configured to operate independently in one of six operation modes: Edge-triggered Capture, Software

Timer, High Speed Output, Frequency Output, 8-Bit Pulse Width Modulator, or 16-Bit Pulse Width Modulator. Each module has Special Function Registers (SFRs) associated with it. These registers are used to exchange data with a module and configure the module's mode of operation. Figure 2.16 shows the PCA Interrupt configuration.

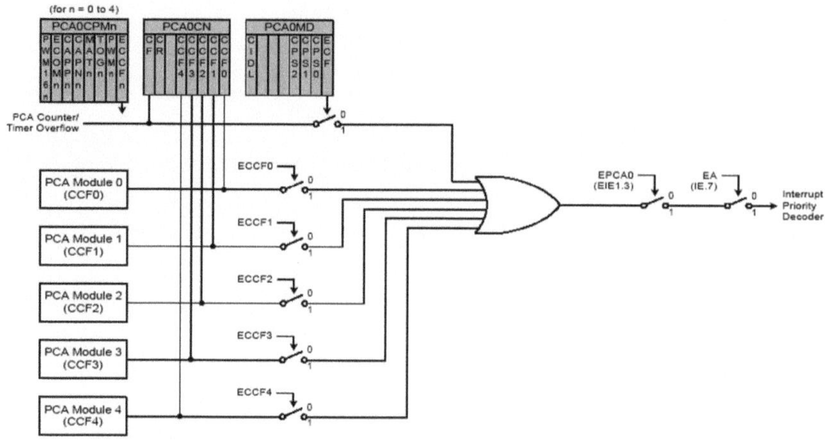

Fig. 2.16 Block diagram of the PCA interrupt structure

Table 2.6 PCA0CPMn register settings

PWM16	ECOM	CAPP	CAPN	MAT	TOG	PWM	ECCF	Operation Mode
X	X	1	0	0	0	0	X	Capture triggered by positive edge on CEXn
X	X	0	1	0	0	0	X	Capture triggered by negative edge on CEXn
X	X	1	1	0	0	0	X	Capture triggered by transition on CEXn
X	1	0	0	1	0	0	X	Software Timer
X	1	0	0	1	1	0	X	High Speed Output
X	1	0	0	X	1	1	X	Frequency Output
0	1	0	0	X	0	1	X	8-Bit Pulse Width Modulator
1	1	0	0	X	0	1	X	16-Bit Pulse Width Modulator

Table 2.6 summarizes the bit settings in the PCA0CPMn registers used to select the PCA0 capture/compare module's operating modes. Setting the ECCFn bit in a PCA0CPMn register enables the module's CCFn interrupt. PCA0 interrupts must be globally enabled before individual CCFn interrupts are recognized. PCA0 interrupts are globally enabled by setting the EA bit (IE.7) and the EPCA0 bit (EIE1.3) to logic 1.

2.11.3 Edge-Triggered Capture Mode

In this mode, a valid transition on the CEXn pin causes PCA0 to capture the value of the PCA0 counter/timer (PCA0L and PCA0H) and load it into the corresponding module's 16-bit capture/compare register (PCA0CPLn and PCA0CPHn).

The CAPPn and CAPNn bits in the PCA0CPMn register are used to select the type of transition that triggers the capture.

- Capture is triggered on low-to-high transition (positive edge) is selected when CAPPn=1 and CAPNn=0
- Capture is triggered on high-to-low transition (negative edge) is selected when CAPPn=0 and CAPNn=1
- Capture is triggered on either transition (positive or negative edge) is selected when CAPPn=1 and CAPNn=1

When a capture occurs, the Capture/Compare Flag (CCFn) in PCA0CN is set to logic 1 and an interrupt request is generated if CCF interrupts have been enabled. The CCFn bit is not automatically cleared by hardware when the CPU vectors to the interrupt service routine, and must be cleared by software.

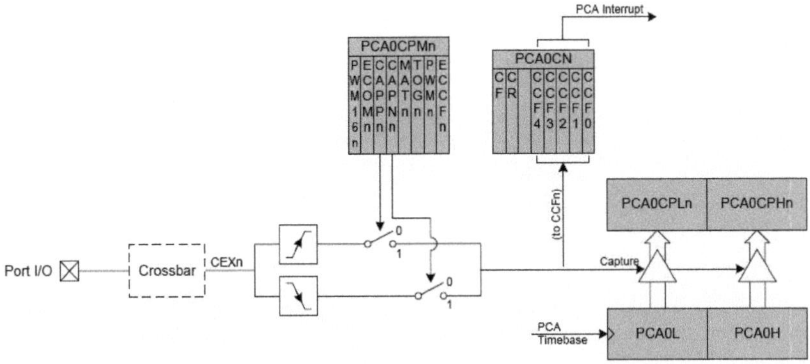

Fig. 2.17 PCA capture mode diagram

2.11.4 Software Timer (Compare) Mode

In Software Timer mode, the PCA0 counter/timer (PCA0L and PCA0H) is compared to the module's 16-bit capture/compare register (PCA0CPHn and PCA0CPLn). When a match occurs, the Capture/Compare Flag (CCFn) in PCA0CN is set to logic 1 and an interrupt request is generated if CCF interrupts are enabled. The CCFn bit is not automatically cleared by hardware when the CPU vectors to the interrupt service routine, and must be cleared by software. Setting the ECOMn and MATn bits in the PCA0CPMn register enables Software Timer (Compare) mode.

When writing a 16-bit value to the PCA0 Capture/Compare registers, the low byte should always be written first. Writing to PCA0CPLn clears the ECOMn bit to ‘0’; writing to PCA0CPHn sets ECOMn to ‘1’. This ensures that the 16-bit comparator is enabled only when the entire 16-bit data has been written to the PCA0 Capture/Compare registers. Figure 2.18 shows the PCA software Timer (Compare) Mode Diagram.

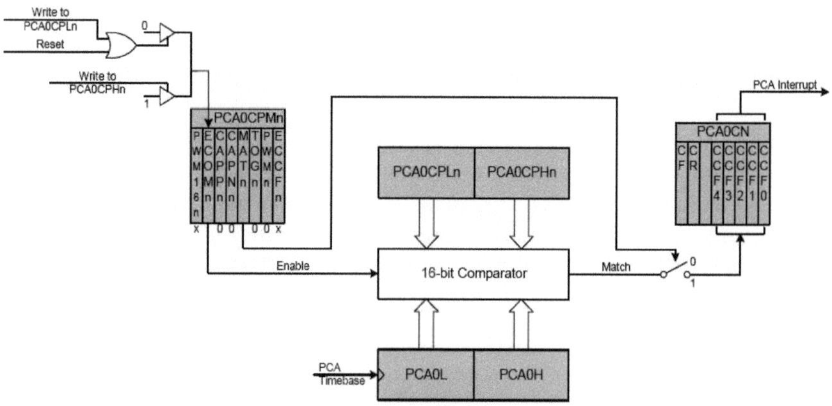

Fig. 2.18 PCA software timer (compare) mode diagram

2.11.5 High Speed Output Mode

In High Speed Output mode, a module’s associated CEXn pin is toggled each time a match occurs between the PCA Counter (PCA0L and PCA0H) and the module's 16-bit capture/compare register (PCA0CPHn and PCA0CPLn). Setting the TOGn, MATn, and ECOMn bits in the PCA0CPMn register enables the High-Speed Output mode.

When writing a 16-bit value to the PCA0 Capture/Compare registers, the low byte should always be written first. Writing to PCA0CPLn clears the ECOMn bit to ‘0’; writing to PCA0CPHn sets ECOMn to ‘1’. This ensures that the 16-bit comparator is enabled only when the entire 16-bit data has been written to the PCA0 Capture/Compare registers. Figure 2.19 shows the PCA High Speed Output Mode Diagram.

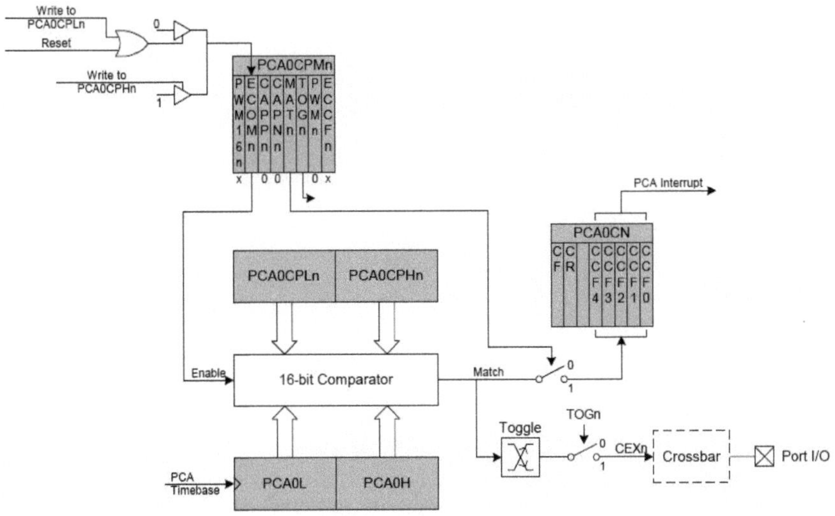

Fig. 2.19 PCA high speed output mode diagram

2.11.6 *Frequency Output Mode*

The Frequency Output Mode produces a programmable-frequency square wave on the module’s associated CEXn pin. The capture/compare register high byte (PCA0CPHn) holds the number of PCA clocks to count before the output is toggled. The frequency of the square wave is defined by the following equation:

$$F_{CEXn} = \frac{F_{PCA}}{2 \times PCA0CPHn}$$

A value of 0x00 in the PCA0CPHn register is equal to 256 for this equation. F_{PCA} is the frequency of the clock selected by the CPS2-0 bits in the PCA mode register (PCA0MD). The lower byte of the capture/compare register (PCA0CPLn) is compared to the PCA0 counter low byte (PCA0L); on a match, CEXn is toggled and the offset held in the high byte is added to the matched value in PCA0CPLn.

Frequency Output Mode is enabled by setting the ECOMn, TOGn, and PWMn bits in the PCA0CPMn register. Figure 2.20 shows the PCA Frequency Output Mode Diagram.

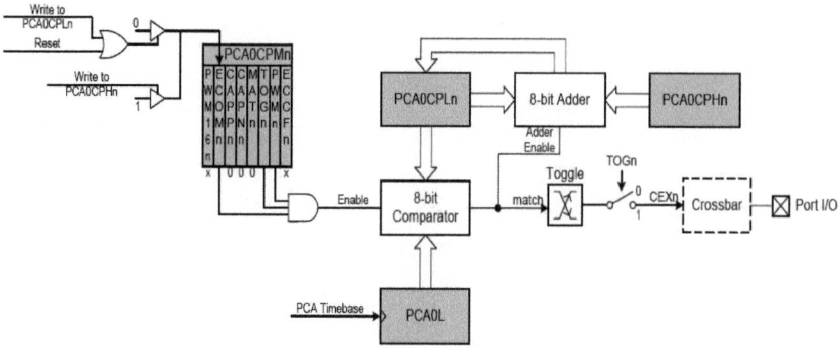

Fig. 2.20 PCA frequency output mode diagram

2.11.7 8-Bit Pulse Width Modulator Mode

Each module can be used independently to generate pulse width modulated (PWM) outputs on its associated CEXn pin. The frequency of the output is dependent on the timebase for the PCA0 counter/timer. The duty cycle of the PWM output signal is varied using the module's PCA0CPLn capture/compare register. When the value in the low byte of the PCA0 counter/timer (PCA0L) is equal to the value in PCA0CPLn, the output on the CEXn pin will be asserted high. When the count value in PCA0L overflows, the CEXn output will be asserted and PCA0CPLn is reloaded automatically with the value stored in the counter/timer's high byte (PCA0CPHn) without software intervention. Setting the ECOMn and PWMn bits in the PCA0CPMn register enables 8-Bit Pulse Width Modulator mode.

The duty cycle for 8-Bit PWM Mode is given by the following equation:

$$DutyCycle = \frac{(256 - PCA0CPHn)}{256}$$

Using the above equation, the largest duty cycle is 100% (PCA0CPHn = 0), and the smallest duty cycle is 0.39% (PCA0CPHn = 0xFF). A 0% duty cycle may be generated by clearing the ECOMn bit to ‘0’.

When writing a 16-bit value to the PCA0 Capture/Compare registers, the low byte should always be written first. Writing to PCA0CPLn clears the ECOMn bit to ‘0’; writing to PCA0CPHn sets ECOMn to ‘1’. This ensures that the 8-bit comparator is enabled only when both the capture/compare registers (PCA0CPLn and PCA0CPHn) are loaded and ready for operation. Figure 2.21 shows the PCA 8-bit PWM Mode Diagram.

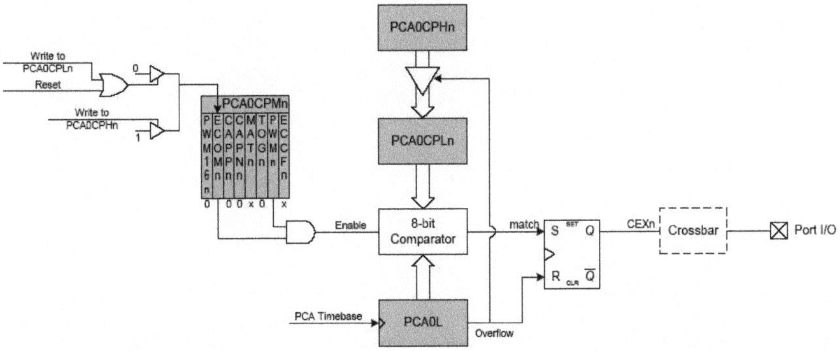

Fig. 2.21 PCA 8-bit PWM mode diagram

The crossbar needs to be programmed to make the two pins, CEX0 and CEX1, available for PWM output. The following code segment configures the crossbar and the GPIO. It enables UART0 and routes CEX0 and CEX1 to two port pins. Thus CEX0 is at pin P0.2 and CEX1 is at pin P0.3.

```
void Init_Port(void)
{
      XBR0 = 0x14;     //-- 00010100: UART0 Enabled, CEX0 and
                       //-- CEX1 routed to 2 port pins
      XBR1 = 0x00;
      XBR2 = 0xC0;     //-- 11000000: Enable Crossbar and
                       //-- disable weak pull-ups

      P0MDOUT |= 0x01;         //-- TX is push-pull
      P0MDOUT &= ~0x02;        //-- RX is open-drain/hi-z
      P0 |= 0x03;              //-- Make sure latches are 1
      P0MDOUT |= 0x0C;         //-- CEX0 and CEX1 are push-pull
}
```

The following program shows how to configure the PCA0 to generate a PWM signal in the 8-bit PWM Mode. SYSCLK used is 22.1184 MHz and the PCA Timebase is SYSCLK/4. PCA Module 0 generates a PWM of 50% duty cycle (at CEX0) while PCA Module 1 generates a PWM (at CEX1) of 75% duty cycle. The frequency of the PWM signal is (22118400/4/256 Hz = 21.6 KHz).

```
void Init_PCA0(void)           //-- Configure the PCA0
{
        PCA0MD = 0x03; //-- 00000011: Use SYSCLK/4 as timebase,
                       //-- enable counter overflow Interrupt
        PCA0CPM0  =  0x42;//--  01000010:  use  PCA  Module  0  for
                        //-- 8-bit PWM generation
        PCA0CPM1  =  0x42;//--  01000010:  use  PCA  Module  1  for
                        //-- 8-bit PWM generation

        PCA0L = 0;

        PCA0CPL0 = 0;
        PCA0CPH0 = 0x80;       //-- 50% duty cycle
        PCA0CPL1 = 0;
        PCA0CPH1 = 0x40;       //-- 75% duty cycle

        PCA0CN = 0x40; // 01000000: Enable PCA0 Counter (CR = 1)
}
```

2.11.8 16-Bit Pulse Width Modulator Mode

Each PCA0 module may also be operated in 16-Bit PWM mode. In this mode, the 16-bit capture/compare module defines the number of PCA0 clocks for the low time of the PWM signal. When the PCA0 counter matches the module contents, the output on CEXn is asserted high; when the counter overflows, CEXn is asserted low. To output a varying duty cycle, new value writes should be synchronized with PCA0 CCFn match interrupts. For a varying duty cycle, CCFn should also be set to logic 1 to enable match interrupts.

The duty cycle for 16-Bit PWM Mode is given by the following equation:

$$DutyCycle = \frac{(65536 - PCA0CPn)}{65536}$$

Using the above equation the largest duty cycle is 100% (PCA0CPn = 0), and the smallest duty cycle is 0.0015% (PCA0CPn = 0xFFFF). A 0% duty cycle may be generated by clearing the ECOMn bit to '0'. 16-Bit PWM Mode is enabled by setting the ECOMn, PWMn, and PWM16n bits in the PCA0CPMn register. When writing a 16-bit value to the PCA0 Capture/Compare registers, the low byte should always be written first. Writing to PCA0CPLn clears the ECOMn bit to '0'; writing to PCA0CPHn sets ECOMn to '1'. This ensures that the 16-bit comparator is enabled only when the entire 16-bit data has been written to the PCA0 Capture/Compare registers. Figure 2.22 shows the PCA 16-bit PWM Mode Diagram.

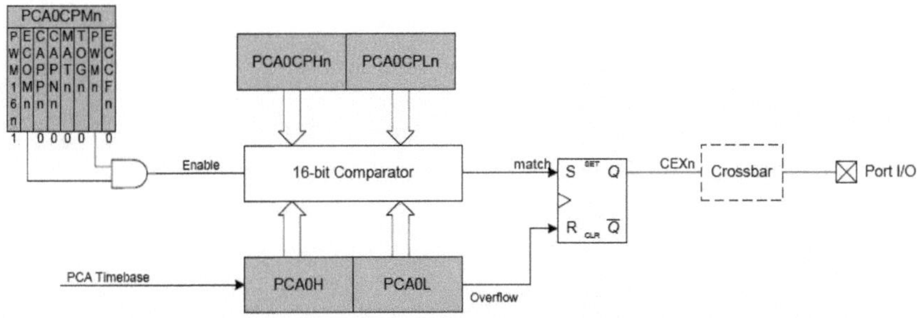

Fig. 2.22 PCA 16-bit PWM mode diagram

The following code segment shows how to configure the PCA0 to generate a PWM signal in the 16-bit PWM Mode. SYSCLK used is 22.1184 MHz and the PCA Timebase is SYSCLK. PCA Module 0 generates a PWM of 50% duty cycle (at CEX0) while PCA Module 1 generates a PWM (at CEX1) of 75% duty cycle. The frequency of the PWM signal is (22118400/65536 = 337.5 Hz). It is assumed that the crossbar has been programmed to make the two pins, CEX0 and CEX1, available for PWM output. While writing the 16-bit value to the PCA0 Capture/Compare registers, the low byte has been written first followed by the high byte.

```
void Init_PCA0(void)          //-- Configure the PCA0
{
      PCA0MD = 0x09; //--  00001001: Use SYSCLK as timebase,
                     //-- enable counter overflow Interrupt
        PCA0CPM0 = 0xC2;//-- 11000010: use PCA Module 0 for
                      //-- 16-bit PWM generation
      PCA0CPM1 = 0xC2;//-- 11000010: use PCA Module 1 for
                      //--16-bit PWM generation

      PCA0L = 0;
      PCA0H = 0;
      PCA0CPL0 = 0x00;
      PCA0CPH0 = 0x80;        //-- 50% duty cycle

      PCA0CPL1 = 0x00;
      PCA0CPH1 = 0x40;        //-- 75% duty cycle

      PCA0CN = 0x40; // 01000000: Enable PCA0 Counter (CR = 1)
}
```

In summary, the benefits of a highly integrated microcontroller include:

1. More efficient circuit implementation and reduced board space
2. Higher system reliability
3. Cost effectiveness

3

C Programming for Silabs C8051F020 Microcontroller

3.1 Introduction

This chapter introduces the Keil™ C compiler for the SiLab C8051F020 board. We assume some familiarity with the C programming language to the level covered by most introductory courses in the C language.

Experienced C programmers, who have little experience with the C8051F020 architecture, should become familiar with the system. The differences in programming the C8051F020 in C, compared to a standard C program, are almost all related to architectural issues. These explanations will be very useful with an understanding of the C8051F020 chip.

The Keil™ C compiler provided with the SiLab C8051F020 board does not come with a floating point library and so the floating point variables and functions should not be used. However if floating point variables are required, a full license for the Keil™ C compiler can be used.

3.2 Register Definitions, Initialization and Startup Code

C is a high level programming language that is portable across many hardware architectures. This means that architecture specific features such as register definitions, initialization and start up code must be made available to a program via the use of libraries and include files.

For the 8051 chip one needs to include the file **reg51.h** or using the SiLab C8051F020-TB development board include the file **c8051f020.h**:

```
#include <reg51.h>
```

or

```
#include < c8051f020.h >
```

G. Sen Gupta and S.C. Mukhopadhyay: Embedded Microcontroller Interfacing, LNEE 65, pp. 49–65.
springerlink.com

These files contain all the definitions of the C8051F020 registers. The standard initialization and startup procedures for the C8051F020 are contained in **startup.a51**. This file is included in the project and will be assembled together with the compiled output of the C program. For custom applications, this startup file might need modification.

3.3 Basic C Program Structure

The following is the basic C program structure; all the programs will have this basic structure.

```
//------------------------------------------------------------
// Basic blank C program that does nothing
// other than disable the watch dog timer
//------------------------------------------------------------

#include <c8051f020.h> // SFR declarations

void main (void)
{
   // disable watchdog timer
      WDTCN = 0xde;
      WDTCN = 0xad;

      while(1);   // Stops program terminating and restarting
}
```

Note: All variables must be declared at the start of a code block. Variables cannot be declared amongst the program statements.

This program can be tested in the SiLab IDE (Integrated Development Environment). One cannot see anything happening on the C8051F020 development board, but it is possible to step through the program using the debugger.

3.4 Programming Memory Models

The C8051F020 processor has 126 Bytes of directly addressable internal memory and up to 64 Kbytes of externally addressable space. The Keil™ C compiler has two main C programming memory models, SMALL and LARGE which are related to these two types of memory. In the SMALL memory model the default storage location is the lower 128 bytes of internal memory while in the LARGE memory model the default storage location is the externally addressed memory.

The default memory model required is selected using the **pragma** *compiler control directive*:

```
#pragma small
int X;
```

Any variable declared in this file (such as the variable *X* above) will be stored in the internal memory of the C8051F020.

The choice of which memory model to use depends on the program, the anticipated stack size and the size of data. If the stack and the data cannot fit in the 128 Bytes of internal memory then the default memory model should be LARGE, otherwise SMALL should be used.

Yet another memory model is the COMPACT memory model. This memory model is not discussed in this chapter. More information on the compact model can be found in the document ***Cx51 Compiler User's Guide for Keil™ Software***.

One can test the different memory models with the SiLab IDE connected to the C8051F020-TB development board. Look at the symbol view after downloading the program and see in which memory addresses the compiler has stored the variables.

3.4.1 Overriding the Default Memory Model

The default memory model can be overridden with the use of Keil™ C programming language extensions that tell the compiler to place the variables in another location. The two main available language extensions are **data** and **xdata**:

```
int data X;
char data Initial;
int xdata Y;
char xdata SInitial;
```

The integer variable *X* and character variable *Initial* are stored in the first 128 bytes of internal memory while the integer variable *Y* and character variable *SInitial* are stored in the external memory overriding any default memory model.

Constant variables can be stored in the read-only code section of the C8051F020 using the **code** language extension:

```
const char code CR=0xDE;
```

In general, access to the internal memory is the fastest, so frequently used data should be stored here while less frequently used data should be stored on the external memory.

The memory storage related language extensions, **bdata,** and associated data types **bit**, **sbit**, **sfr** and **sfr16** will be discussed in the following sections. Additional memory storage language extensions including, **pdata** and **idata,** are not discussed in this chapter; refer to the document ***Cx51 Compiler User's Guide for Keil™ Software*** for information on this.

3.4.2 Bit-Valued Data

Bit-valued data and bit-addressable data must be stored in the bit-addressable memory space on the C8051F020 (0x20 to 0x2F). This means that bit- valued data and bit-addressable data must be labeled as such using the **bit**, **sbit** and **bdata**.

Bit-addressable data must be identified with the **bdata** language extension:

```
int bdata X;
```

The integer variable X declared above is bit-addressable.

Any bit valued data must be given the **bit** data type, this is not a standard C data type:

```
bit flag;
```

The bit-valued data *flag* is declared as above.

The **sbit** data type is used to declare variables that access a particular bit field of a previously declared bit-addressable variable.

```
bdata X;
sbit X7flag = X^7;   /* bit 7 of X */
```

X7flag declared above is a variable that references bit 7 of the integer variable *X*.

You cannot declare a bit pointer or an array of bits.

The bit valued data segment is 16 bytes or 128 bits in size, so this limits the amount of bit-valued data that a program can use.

3.4.3 Special Function Registers

As can be seen in the include files **c8051f020.h** or **reg51.h**, the special function registers are declared as a **sfr** data type in Keil™ C. The value in the declaration specifies the memory location of the register:

```
/*  BYTE Register  */
sfr P0   = 0x80;
sfr P1   = 0x90;
```

Extensions of the 8051 often have the low byte of a 16 bit register preceding the high byte. In this scenario it is possible to declare a 16 bit special function register, **sfr16**, giving the address of the low byte:

```
sfr16 TMR3RL = 0x92;          // Timer3 reload value
sfr16 TMR3   = 0x94;          // Timer3 counter
```

The memory location of the register used in the declaration must be a constant rather than a variable or expression.

3.4.4 Locating Variables at Absolute Addresses

Variables can be located at a specific memory location using the **_at_** language extension:

```
int X _at_  0x40;
```

The above statement locates the integer *X* at the memory location 0x40.

The **_at_** language extension can not be used to locate bit addressable data.

3.5 C Language Operators and Control Structures

C language is a structured programming language that provides **sequence**, **selection** and **repetition** language constructs to control the flow of a program. The sequence in which the program statements execute is one after another within a code block. Selection of different code blocks is determined by evaluating **if** and **else if** statements (as well as **switch-case** statements) while repetition is determined by the evaluation of **for** loop or **while** loop constructs.

3.5.1 Relational Operators

Relational operators compare data and the outcome is either True or False. The **if** statements, **for** loops and **while** loops can make use of C relational operators. These are summarized in Table 3.1.

Table 3.1 Relational Operators

Operator	Description
==	Equal to
!=	Not Equal to
<	Less than
>	Greater than
<=	Less than or equal to
>=	Greater than or equal to

3.5.2 Logical Operators

Logical operators operate on Boolean data (True and False) and the outcome is also Boolean. The logical operators are summarized in Table 3.2.

Table 3.2 Logical Operators

<table>
<tr><th>Operator</th><th>Description</th></tr>
<tr><td>&&</td><td>Logical AND</td></tr>
<tr><td>||</td><td>Logical OR</td></tr>
<tr><td>!</td><td>Logical NOT</td></tr>
</table>

3.5.3 Bitwise Logical Operators

As well as the Logical operators that operate on integer or character data, the C language also has bitwise logical operators. These are summarized in Table 3.3.

Table 3.3 Bit valued logical operators

<table>
<tr><th>Operator</th><th>Description</th></tr>
<tr><td>&</td><td>Bitwise AND</td></tr>
<tr><td>|</td><td>Bitwise OR</td></tr>
<tr><td>~</td><td>Bitwise NOT</td></tr>
<tr><td>^</td><td>Bitwise XOR</td></tr>
</table>

Bitwise logical operators operate on each bit of the variables individually.

Example:

```
X = 0x40 | 0x21;
```

The above statement will assign the value 0x61 to the variable X.

```
0x40        0100 0000
0x21        0010 0001
            ---------   Bitwise Logical OR
0x61        0110 0001
```

3.5.4 Compound Operators

C language provides short cut bitwise operators acting on a single variable similar to the +=, -=, /= and *= operators. These are summarized in Tables 3.4 and 3.5.

Table 3.4 Compound Arithmetic Operators

Operator	Description	Example	Equivalent
+=	Add to variable	X += 2	X=X + 2
-=	Subtract from variable	X -= 1	X=X - 1
/=	Divide variable	X /= 2	X=X / 2
*=	Multiply variable	X *= 4	X=X * 4

Table 3.5 Compound Bitwise Operators

<table>
<tr><th>Operator</th><th>Description</th><th>Example</th><th>Equivalent</th></tr>
<tr><td>&=</td><td>Bitwise And with variable</td><td>X &= 0x00FF</td><td>X=X & 0x00FF</td></tr>
<tr><td>|=</td><td>Bitwise Or with variable</td><td>X |= 0x0080</td><td>X=X | 0x0080</td></tr>
<tr><td>^=</td><td>Bitwise XOR with variable</td><td>X ^= 0x07A0</td><td>X=X | 0x07A0</td></tr>
</table>

Initializing Crossbar and GPIO Ports

We can initialize the crossbar and GPIO ports using the C bitwise operators.

```
//-- Configures the Crossbar and GPIO ports
  XBR2 = 0x40;        //-- Enable Crossbar and weak
                      //    pull-ups (globally)
  P1MDOUT |= 0x40;    //-- Enable P1.6 as push-pull output
```

3.5.5 Making Choices

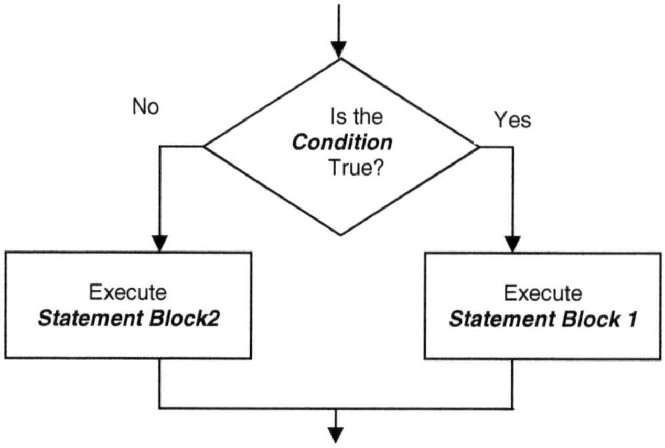

Fig. 3.1 Flow chart for a ***selection*** construct

Choices are made in the C language using an **if else** statement.

```
if (x > 10)              //-- the condition is true
    { y=y+1; }           //-- Execute statement block 1
else                     //-- the condition is false
    { y=y-1; }           //-- Execute statement block 2
```

When the *Condition* is evaluated as True the first block is executed and if the *Condition* evaluates as being False the second block is executed.

More conditions can be created using a sequence of **if** and **else if** statements.

```
if (x > 10)
        { y=y+1; }
   else if (x > 0)
              { y=y-1; }
        else
              { y=y-2; }
```

In some situations, when there is a list of integer or character choices a **switch-case** statement can be used.

```
switch (x)
{
 case 5:
     y=y+2; break;
 case 4: case 3:
     y=y+1; break;
 case 2: case 1:
     y=y-1; break;
 default:
     y=y-2; break;
}
```

When the variable *x* in the **switch** statement matches one of the case statements, that block is executed. Only when the **break** statement is reached does the flow of control break out of the switch statement. The default block is executed when there are no matches with any of the case statements.

If the **break** statements are missing from the **switch-case** statement then the flow will continue within the **switch-case** block until a **break** statement or the end of the **switch-case** block is reached.

3.5.6 *Repetition*

Numeric repetition of a code block for a fixed set of times is achieved using a for loop construct.

```
int i;
int sum=0;
for( i = 0; i<10; i++)
{
     sum = sum + i;
}
```

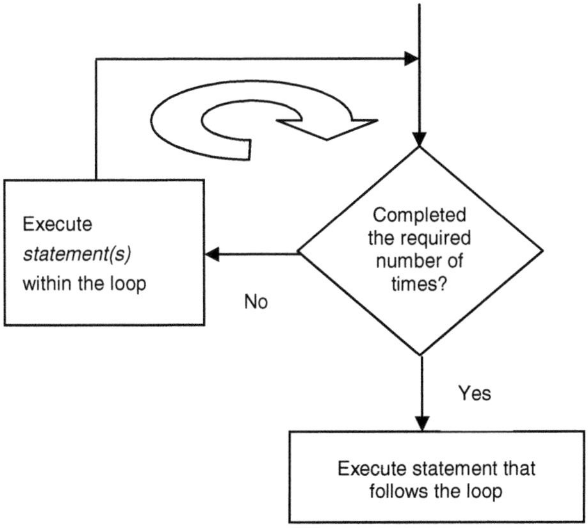

Fig. 3.2 Flow chart for a **for** loop

When the looping required is not determined by a fixed number of counts but by a more complex condition, we normally use the **while** loop construct to control the process.

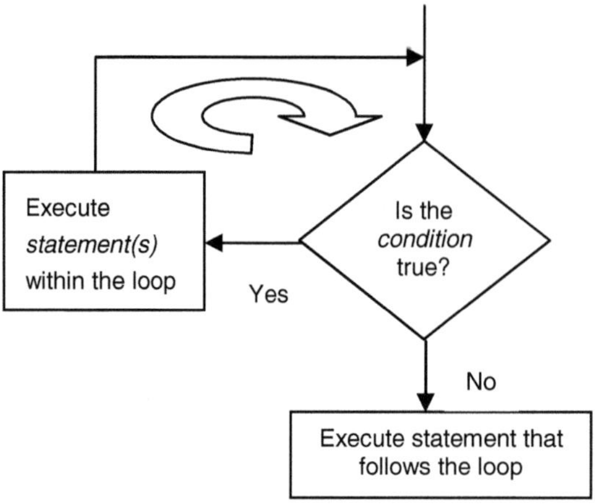

Fig. 3.3 Flow chart for a **while** loop

The **while** loop repeats the loop while the condition specified is true.

3.5.7 Waiting for Events

We can use a **while** loop to wait for the crystal oscillator valid flag to be set.

```
//-- wait till XTLVLD pin is set
while ( !(OSCXCN & 0x80) );
```

3.5.8 Early Exits

When executing a code block or a loop, sometimes it is necessary to exit the current code block. The C language provides several mechanisms to do this.

The **break** statement will move the flow of control outside the end of the current loop.

```
int i;
int sum=0;
for( i = 0; i<10; i++)
{
        sum = sum + i;
        if (sum > 25) break;
}
```

The **continue** statement skips the remaining code in the current loop, but continues from the start of the code block of the loop (after incrementing and checking that the loop should not terminate)

```
int i;
int sum=0;
for( i = 0; i<10; i++)
{
      if (i == 5) continue;
      sum = sum + i;
}
```

3.6 Functions

Functions in C are declared using the return data type, the data type of the parameters and the body of the function.

```
unsigned long square (int x)
{
      return x*x;
}
```

Standard functions in Keil™ C are not re-entrant and so should not be called recursively. This is the case as parameters and local variables are stored in a standard location for all calls to a particular function. This means that recursive calls will corrupt the data passed as arguments to the function as well as the local variables.

A stack, starting straight after the last data stored in internal memory is used to keep track of function calls, but only the return address is stored on the stack, so conserving space. One can see the operation of the stack in the SiLab IDE.

Test the functions using the SiLab IDE connected to the C8051F020 development board. It may be noticed that sometimes the compiler optimizations will result in some variables sharing the same memory address!

3.6.1 Standard Function – Initializing System Clock

A C function to initialize the system clock can be written as follows-

```
void Init_Clock(void)
{
    OSCXCN = 0x67;          //-- 0110 0111b
    //-- External Osc Freq Control Bits (XFCN2-0) set
    //   to 111 because crystal frequency > 3.7 MHz
    //-- Crystal Oscillator Mode (XOSCMD2-0) set to 110

    //-- wait till XTLVLD pin is set
    while ( !(OSCXCN & 0x80) );

    OSCICN = 0x88;          //-- 1000 1000b
    //-- Bit 2 : Internal Osc. disabled (IOSCEN = 0)
    //-- Bit 3 : Uses External Oscillator as System
    //           Clock (CLKSL = 1)
    //-- Bit 7 : Missing Clock Detector Enabled (MSCLKE = 1)
}
```

3.6.2 Memory Model Used for a Function

The memory model used for a function can override the default memory model with the use of the **small**, **compact** or **large** keywords.

```
int square (int x) large
{
     return x*x;
}
```

3.7 Interrupt Functions

The basic 8051 has 5 possible interrupts which are listed in Table 3.6.

Table 3.6 8051 Interrupts

Interrupt No.	Description	Address
0	External INT 0	0x0003
1	Timer/ Counter 0	0x000B
2	External INT 1	0x0013
3	Timer/ Counter 1	0x001B
4	Serial Port	0x0023

The Cx51 has extended these to 22 interrupts to handle additional interrupts provided by manufacturers (see chapter appendix for the full table). An interrupt function is declared using the **interrupt** key word followed by the required interrupt number.

```
int count;

void timer1_ISR (void) interrupt 3
{
    count++;
}
```

Interrupt functions must not take any parameters and not return any parameters. Interrupt functions will be called automatically when the interrupt is generated; they should not be called in normal program code, this will generate a compiler error.

3.7.1 Timer 3 Interrupt Service Routine

One can write a timer 3 Interrupt Service Routine (ISR) that changes the state of an LED depending on whether a switch is pressed-

```
sbit LED = P1^6;        //-- LED at port pin P1.6
int LED_count=0;

//-- This routine changes the state of the LED
//   whenever Timer3 overflows.

void Timer3_ISR (void) interrupt 14
{
  unsigned char P3_input;
  TMR3CN &= ~(0x80);        //-- clear TF3

  P3_input = ~P3;
  if (P3_input & 0x80)      //-- if bit 7 is set,
  {                         //   then switch is pressed
     LED_count++;
     if ( (LED_count % 10) == 0)
     {                       //-- do every 10th count
       LED = ~LED;           //-- change state of LED
       LED_count = 0;
     }
  }
}
```

3.7.2 Disabling Interrupts before Initialization

Before using interrupts (such as the timer interrupts) they should be initialized. Before initialization interrupts should be disabled so that there is no chance that the interrupt service routine is called before initialization is complete.

```
EA = 0;              //-- disable global interrupts
```

When initialization has been completed the interrupts can be enabled.

```
EA = 1;              //-- enable global interrupts
```

3.7.3 Timer 3 Interrupt Initialization

We can put the timer 3 initialization statements within a C function

```
//-- Configure Timer3 to auto-reload and generate
//-- an interrupt at interval specified by <counts>
//-- using SYSCLK/12 as its time base.

void Init_Timer3 (unsigned int counts)
{
    TMR3CN = 0x00; //-- Stop Timer3; Clear TF3;
                   //-- use SYSCLK/12 as timebase

    TMR3RL  = -counts; //-- Init reload values
    TMR3    = 0xffff;  //-- set to reload immediately
    EIE2   |= 0x01;    //-- enable Timer3 interrupts
    TMR3CN |= 0x04;    //-- start Timer3 by setting
                       //   TR3 (TMR3CN.2) to 1
}
```

3.7.4 Register Banks

Normally a function uses the default set of registers. However there are 4 sets of registers available in the C8051F020. The register bank that is currently in use can be changed for a particular function via the **using** Keil™ C language extension.

```
int count;

void timer1 (void) interrupt 3 using 1
{
        count++;
}
```

The register bank specified by the using statement ranges from 0 to 3. The register bank can be specified for normal functions, but are more appropriate for interrupt functions. When no register bank is specified in an interrupt function the state of the registers must be stored on the stack before the interrupt service routine is called. If a new register bank is specified then only the old register bank number needs to be copied to the stack significantly improving the speed of the interrupt service routine.

3.8 Reentrant Functions

Normal Keil™ C functions are not re-entrant. A function must be declared as re-entrant to be able to be called recursively or to be called simultaneously by two or more processes. This capability is often required in real-time applications or in situations when interrupt code and non-interrupt code need to share a function.

```
int fact (int X) reentrant
{
        if ( X==1) { return 1; }
        else { return X*fact(X-1); }
}
```

A re-entrant function stores the local variables and parameters on a simulated stack. The default position of the simulated stack is at the end of internal memory (0xFF). The starting positions of the simulated stack are initialized in **startup.a51** file.

The simulated stack makes use of indirect addressing; this means that when one uses the debugger and watches the values of the variables they will contain the address of the memory location where the variables are stored. One can view the internal RAM (address 0xff and below) to see the parameters and local variable placed on the simulated stack.

3.9 Pointers

Pointers in C are a data type that stores the memory addresses. In standard C the data type of the variable stored at that memory address must also be declared:

```
int * X;
```

3.9.1 A Generic Pointer in Keil™ C

Since there are different types of memory on the C8051F020 processor there are different types of pointers. These are *generic pointers* and *memory specific pointers*. In standard C language we need to declare the correct data type that the pointer points to. In Keil™ C we also need to be mindful of which memory model we are pointing to when we are using memory-specific pointers. Generic pointers remove this restriction, but are less efficient as the compiler needs to store what memory model is being pointed to. This means that a generic pointer takes 3 bytes of storage - 1 byte to store the type of memory model that is pointed to and two bytes to store the address.

```
int * Y;
char * ls;
long * ptr;
```

One may also explicitly specify the memory location that the generic pointer is stored in, to override the default memory model.

```
int * xdata Y;
char * idata ls;
long * data ptr;
```

3.9.2 Memory Specific Pointers

A memory specific pointer points to a specific type of memory. This type of pointer is efficient as the compiler does not need to store the type of memory that is being pointed to. The data type of the variable stored at the memory location must be specified.

```
int xdata * Y;
char data * ls;
long idata * ptr;
```

You may also specify the memory location that the memory-specific pointer is stored in, to override the default memory model.

```
int data * xdata Y;
char xdata * idata ls;
long idata * data ptr;
```

3.10 Summary of Data Types

In Table 3.7, we have summarized the Data Types that are available in the Cx51 compiler. The size of the data variable and the value range is also given.

Table 3.7 Data Types

Data Type	Bits	Bytes	Value Range
bit	1	-	0 to 1
signed char	8	1	-128 to +127
unsigned char	8	1	0 to 255
enum	8/16	1 or 2	-128 to +127 or -32768 to +32767
signed short	16	2	-32768 to +32767
unsigned short	16	2	0 to 65535
signed int	16	2	-32768 to +32767
unsigned int	16	2	0 to 65535
signed long	32	4	-2147483648 to 2147483647
unsigned long	32	4	0 to 4294967295
float	32	4	±1.175494E-38 to ±3.402823E+38
sbit	1	-	0 to 1
sfr	8	1	0 to 255
sfr16	16	2	0 to 65535

A3 Chapter Appendix

Table A.1 Interrupt Summary

Interrupt Source	Interrupt Vector	Priority Order	Pending Flag	Enable Flag	Priority Control
Reset	0000	Top	None	Always Enabled	Always Highest
External Interrupt 0 (/INT0)	0003	0	IE0 (TCON.1)	EX0 (IE.0)	PX0 (IP.0)
Timer 0 Overflow	000B	1	TF0 (TCON.5)	ET0 (IE.1)	PT0 (IP.1)
External Interrupt 1 (/INT1)	0013	2	IE1 (TCON.3)	EX1 (IE.2)	PX1 (IP.2)
Timer 1 Overflow	001B	3	TF1 (TCON.7)	ET1 (IE.3)	PT1 (IP.3)
UART0	0023	4	RI0 (SCON0.0) TI0 (SCON0.1)	ES0 (IE.4)	PS0 (IP.4)
Timer 2 Overflow	002B	5	TF2 (T2CON.7)	ET2 (IE.5)	PT2 (IP.5)
Serial Peripheral Interface	0033	6	SPIF (SPI0CN.7)	ESPI0 (EIE1.0)	PSPI0 (EIP1.0)
SMBus Interface	003B	7	SI (SMB0CN.3)	ESMB0 (EIE1.1)	PSMB0 (EIP1.1)
ADC0 Window Comparator	0043	8	AD0WINT (ADC0CN.2)	EWADC0 (EIE1.2)	PWADC0 (EIP1.2)
Programmable Counter Array	004B	9	CF (PCA0CN.7) CCFn (PCA0CN.n)	EPCA0 (EIE1.3)	PPCA0 (EIP1.3)
Comparator 0 Falling Edge	0053	10	CP0FIF (CPT0CN.4)	ECP0F (EIE1.4)	PCP0F (EIP1.4)
Comparator 0 Rising Edge	005B	11	CP0RIF (CPT0CN.5)	ECP0R (EIE1.5)	PCP0R (EIP1.5)
Comparator 1 Falling Edge	0063	12	CP1FIF (CPT1CN.4)	ECP1F (EIE1.6)	PCP1F (EIP1.6)
Comparator 1 Rising Edge	006B	13	CP1RIF (CPT1CN.5)	ECP1R (EIE1.7)	PCP1F (EIP1.7)
Timer 3 Overflow	0073	14	TF3 (TMR3CN.7)	ET3 (EIE2.0)	PT3 (EIP2.0)
ADC0 End of Conversion	007B	15	AD0INT (ADC0CN.5)	EADC0 (EIE2.1)	PADC0 (EIP2.1)
Timer 4 Overflow	0083	16	TF4 (T4CON.7)	ET4 (EIE2.2)	PT4 (EIP2.2)
ADC1 End of Conversion	008B	17	AD1INT (ADC1CN.5)	EADC1 (EIE2.3)	PADC1 (EIP2.3)
External Interrupt 6	0093	18	IE6 (PRT3IF.5)	EX6 (EIE2.4)	PX6 (EIP2.4)
External Interrupt 7	009B	19	IE7 (PRT3IF.6)	EX7 (EIE2.5)	PX7 (EIP2.5)
UART1	00A3	20	RI1 (SCON1.0) TI1 (SCON1.1)	ES1 (EIE2.6)	PS1 (EIP2.6)
External Crystal OSC Ready	00AB	21	XTLVLD (OSCXCN.7)	EXVLD (EIE2.7)	PXVLD (EIP2.7)

4

Design Issues of Microcontroller Interfacing

4.1 Introduction

In this chapter design issues of interfacing microcontrollers to some common electronic circuits have been discussed. These important issues are critical to the success of a project. Open collector configuration, loading problems, microcontroller's cross-bar configuration, driving output load etc are a few of the issues discussed.

4.2 Open-Collector Configuration

In many situations the output of a circuit is in an open-collector configuration. The term open-collector usually refers to a transistor output. In this configuration the collector of the transistor is kept open i.e., it is not connected to the positive supply, as shown in figure 4.1. Usually the transistor operates in cut-off or saturation regions i.e., as a switch. For proper functioning, the collector of the transistor should be connected to a positive supply through a pull-up resistor to complete the circuit. This provides an advantage to the designer as the pull-up resistor can be connected to a range of different voltages. The voltage level should be above the transistor saturation level.

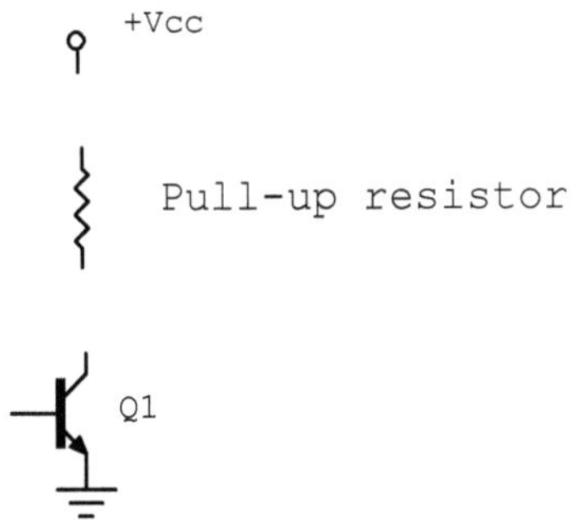

Fig. 4.1 Open-collector configuration

G. Sen Gupta and S.C. Mukhopadhyay: Embedded Microcontroller Interfacing, LNEE 65, pp. 67–82.
springerlink.com

Another advantage of open-collector configuration is to interface devices with different voltage levels. If instead of a bipolar junction transistor (BJT) a MOSFET is used, the term open-drain is commonly used.

The choice of voltage level depends on the application and it must be within the allowable limit of the transistor. The value of the pull-up resistor is to be properly selected so that the current through the transistor doesn't exceed the allowable limit of the transistor. Since the transistor operates in saturation, the current through the pull-up resistor, R, as well as through the transistor is approximately equal to V_{cc}/R neglecting the collector to emitter voltage drop. Typical value of the collector current is in the range of 10mA to 100mA. If the current exceeds the allowable limit, the transistor may burn out.

In many situations instead of a pull-up resistor, different types of loads, such as power relay, solenoid, motor, coil, or incandescent lamp may be used. Some loads (power relay, solenoids, motors, coils) are inductive in nature. Usually an inductive load generates a very high voltage spike when the switch is turned off. The designer should be very careful while those types of loads are used. If it is unavoidable, the transistors must be protected from transient over-voltages. The use of transient suppression components, RC filter, and free-wheeling diode parallel to the load is useful. Also a snubber or a zener diode across the transistor may be helpful. This is critical, as a single transient pulse may damage the transistor.

In case of capacitive load, though rare in practice, it must be assured that the inrush current does not exceed the maximum current rating of the transistor.

If the load is an incandescent lamp, care must be taken because it has a very high start-up current. The filament glows and the resistance of the lamp settles to a steady-state value. Usually it is not recommended to use incandescent lamp for open-collector output; rather LED is to be used.

4.3 Protection of Microcontroller from Over-Voltage

In this section we will discuss a few tips to protect electronic circuits, especially the microcontroller, while the external signals are interfaced. The damage is caused due to over-overvoltage. With the help of some simple protection devices, it is possible to increase the EMI (Electro Magnetic Interference) and ESD (Electro Static Discharge) immunity level of the complete circuit and system. Usually all semiconductor devices, mainly ICs, contain internal protection circuits but it is not practical to incorporate large protection devices. The external protection devices provide a higher level of surge protection. Though the knowledge of internal surge protection circuit may be helpful in selecting an external protection device with an appropriate power rating and turn-on voltages, the data sheets usually do not disclose the details of internal protection circuits.

The main function of an external protection device is to limit the current through an IC or microcontroller by reducing the magnitude of surge voltage. It is expected that the protection device will turn on before the internal circuit turns on and absorb the entire energy of the surge pulse. The location of the protection device is very important to determine whether the majority of the surge energy is absorbed by the external protection circuit. The layout of the printed circuit board (PCB) is also very important for proper operation of the protection circuit.

Figure 4.2 shows a protection circuit with a diode array. A value of 0.7 V can be used to estimate the turn-on voltage of the external switching circuit. So the voltage level at input signal will be restricted to -0.7V to V_{DD}+0.7 V. The diodes should respond very quickly for proper operation.

In many situations schottky diodes of typical turn on voltage of 0.3 V may be used as is shown in figure 4.3. In this case the voltage level at the input signal is restricted to -0.3V to V_{DD}+0.3 V.

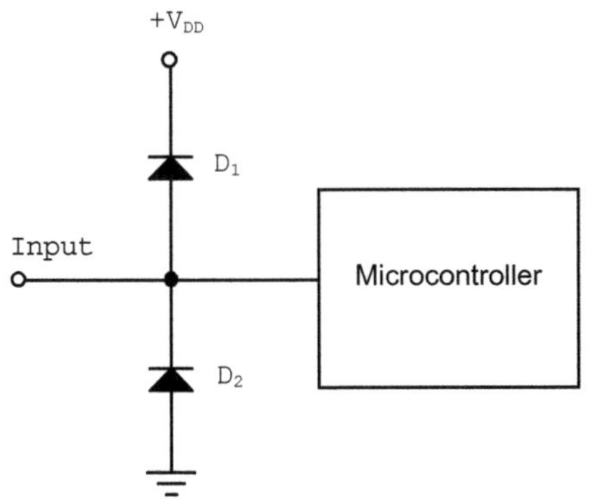

Fig. 4.2 A diode array for transient voltage protection

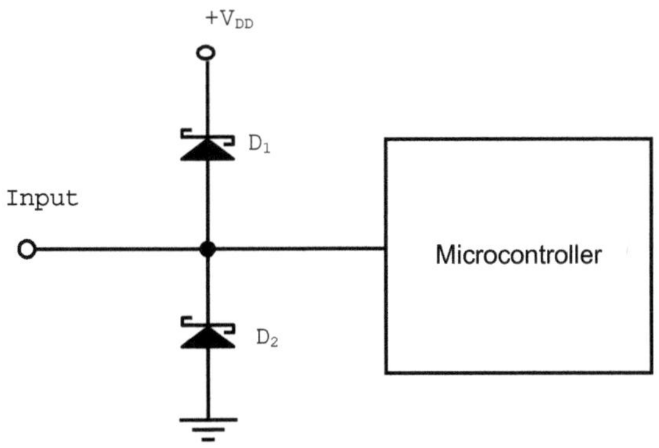

Fig. 4.3 Schottky diode array based transient voltage protection

The circuit shown in figure 4.4 is another way of dealing with the problem. The series resistance ensures that the majority of the surge energy will be dissipated by the external protection circuit. The current, I_2, which flows to the microcontroller, is relatively very small compared to the current I_1, which flows to the protective device.

The protection using diode arrays steer the surge current into the power supply rails where the energy of the transient voltage pulse is dissipated. If the energy associated with the transient voltage pulse is considerably large, it may affect the voltage level of the power supply rails. A decoupling capacitor, along with avalanche diode, can be used to improve the load regulation of a power supply during a surge effect. A high frequency ceramic capacitor of approximately 0.01 to 0.1 μF across the power pins reduces the effect of the surge pulse.

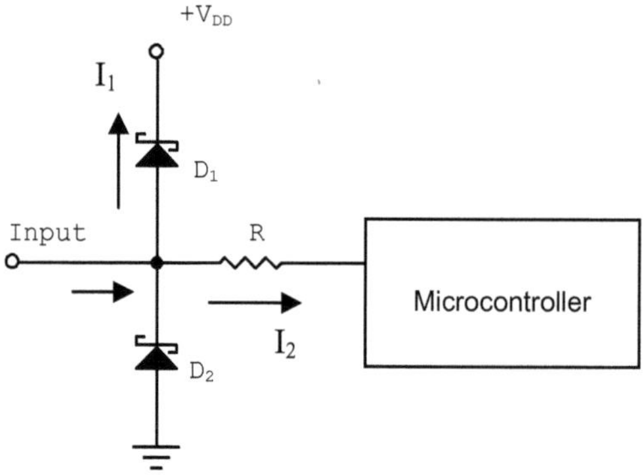

Fig. 4.4 Schottky diode array based transient voltage protection along with series resistance

If an avalanche diode with a breakdown voltage slightly higher than V_{DD} is used across the supply, an additional surge protection can be achieved. The schematic representation is shown in figure 4.5.

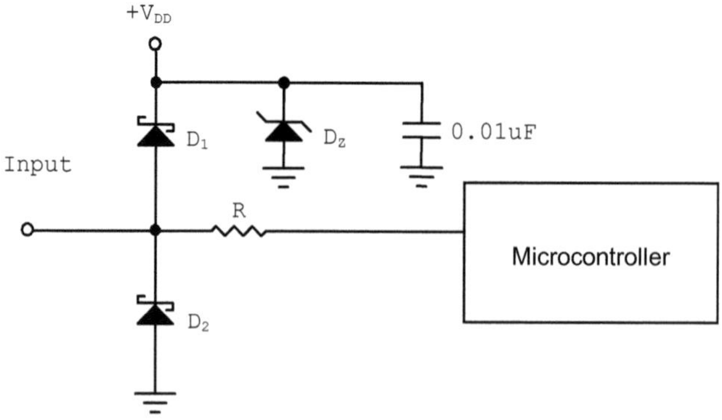

Fig. 4.5 Transient voltage protection along with decoupling capacitor and avalanche diode

4.4 Switching Inductive Load and Diode Protection

In many situations the microcontrollers are used with inductive loads such as motors, solenoids, and relays which generate switching transient voltages of many times the steady state value. Let us see what happens if we open a switch that is providing current to an inductor. The inductor voltage and current are related by:

$$V_L = L\frac{di_L}{dt}$$

where V_L is the voltage across the inductor and i_L is the current through the inductor.

It is not possible to turn off the current suddenly as this would mean a very large voltage (ideally infinite) V_L would appear across the inductor's terminals. For example, turning off a 12 V solenoid can easily create a negative spike of around 300 volts. If the device can withstand this high voltage, the device or the system can survive. In the worst case, the switching transient voltage can destroy the microcontroller and semiconductor devices. In some situation, the transient voltage can cause program failures and flash memory corruption. In the case of high current, large inductance devices, the spike need not even be directly connected to the microcontroller to cause the damage or program failure. Microcontrollers or systems damaged from inductive spikes are considered to be abused and are not eligible for warranty repair.

Figure 4.6a shows the circuit diagram of a common circuit generally used to switch current through an inductive load. The switch is initially closed (ON) and the inductor (relay, motor, solenoid etc.) is carrying the full current. When the switch is tuned off, the inductor tries to continue the flow of current in the direction as shown by the arrow in figure 4.6a. In the absence of any path, the current will generate a huge voltage spike which may damage the switch. Even if the switch is not damaged, it definitely shortens the life of the switch and generates electromagnetic interference that may affect the nearby neighbouring circuits.

A simple solution to avoid this type of problem is to provide a diode across the inductor as is shown in figure 4.6b. The diode doesn't conduct while the switch is ON as it is reverse biased. When the switch is tuned off, the inductive spike makes the diode forward biased and the current flows through the diode. The diode should be able to handle the inductor current which was flowing just before the switch goes to OFF state. In many situations a diode like 1N4004 is fine for nearly all cases. The diode is also known as a freewheeling diode, flyback diode, suppressor diode or catch diode.

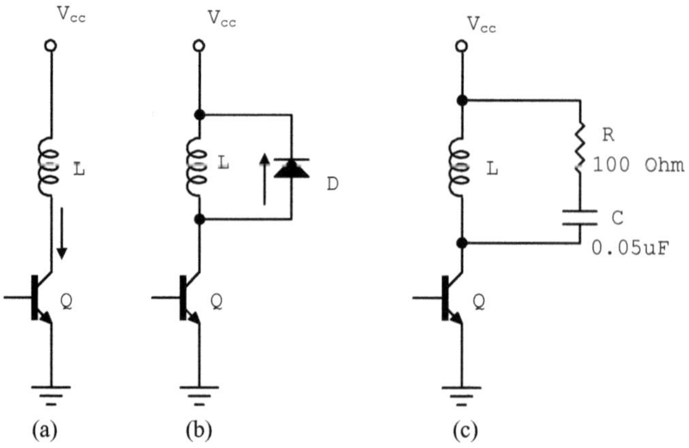

Fig. 4.6 Switching transient overvoltage and protection circuit

While the current 'free-wheels' through the diode the decay of the inductor current is decided by the voltage drop across the diode. This may take a very long time in some situation. In order to decay the inductor current very quickly a resistance can be used in series with the diode.

While dealing with alternating supply voltage a RC snubber, as is shown in figure 4.6c, is to be used in place of the diode. Also in many applications, it is recommended to use a varistor in parallel with the load. The rating of the varistor voltage should be about 1.5 times the peak-to-peak steady-state voltage of the load.

It is recommended not to use the microcontroller's ground or power conductors to carry inductively switched loads. It is highly recommended to route all such conductors directly to and from the power supply and should be located as far away from the controller as possible. The use of a separate power source for large inductive loads is strongly recommended. It is also a good practice to use a separate enclosure for the microcontroller to shield it from the electromagnetic interference.

4.5 Potential Divider for Feedback Voltage

In many control engineering problems a feedback signal is required to be fed to the microcontroller for taking necessary control action. The feedback voltage is derived by using a suitable potential divider. The use of voltage or potential divider is very common in electronic circuits. It is a circuit which provides a fraction of the given voltage. The choice of resistances of the potential divider as well as an understanding of the operation of the circuit from Thevenin's equivalent theorem is important to achieve the desired output. A typical example using a switched mode power supply (SMPS) has been considered to explain the issue.

A boost converter with the following specification has been considered. The schematic diagram is shown in figure 4.7.

Input voltage:	4V
Nominal output voltage:	8V
Nominal output current:	100 mA
Nominal operating frequency:	100 KHz

The output voltage of a boost converter is related to the input voltage of it by-

$$V_{out} = \frac{V_{in}}{1 - D}$$

where D is the nominal duty ratio. The nominal duty ratio for this case is 0.5.

The feedback voltage is taken with the help of the potential divider formed by the resistances R_1 and R_2 and is given by-

$$V_{feedback} = \frac{R_2}{R_1 + R_2} V_{out}$$

Ideally, V_{out} is expected to be 8V. However, this output voltage can change depending on the output condition. With the help of some corrective action the output is kept constant.

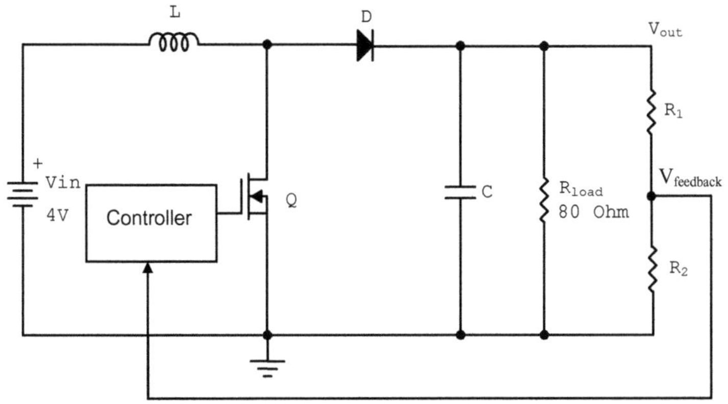

Fig. 4.7 Boost converter and feedback voltage

Usually the signal taken from the potential divider is fed to an electronic circuit. So the situation, as is shown in figure 4.8a, is not common. The open circuit voltage at the junction point is given by the expression of $V_{feedback}$. Assuming the feedback voltage is feeding an equivalent impedance, the actual situation is as shown in figure 4.8b, in which the feedback voltage is basically feeding a load having an external resistance of R_{ext}.

The situation can be represented with the help of equivalent Thevenin's circuit as shown in figure 4.8c.

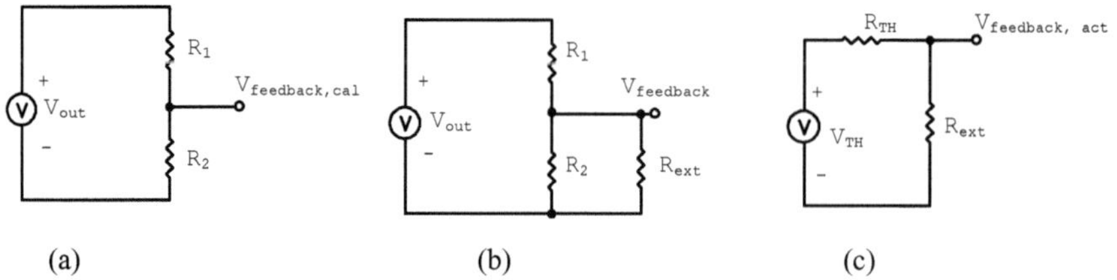

Fig. 4.8 Analysis of potential divider circuit

The equivalent Thevenin's voltage and resistance can be calculated as-

$$V_{Th} = \frac{R_2}{R_1 + R_2} V_{out} \quad \text{and} \quad R_{TH} = \frac{R_1 R_2}{R_1 + R_2}$$

Let us take $R_2 = n * R_1$ so that theoretically the feedback voltage will remain constant.

$$V_{feedback,cal} = \frac{R_2}{R_1 + R_2} V_{out} = \frac{n}{n+1} V_{out}$$

$$V_{Th} = \frac{R_2}{R_1 + R_2} V_{out} = \frac{n}{n+1} V_{out} \quad \text{and}$$

$$R_{TH} = \frac{R_1 R_2}{R_1 + R_2} = \frac{n}{n+1} R_1$$

So, the actual feedback voltage is given by-

$$V_{feedback,act} = \frac{R_{ext}}{R_{ext} + R_{TH}} V_{TH} = \frac{R_{ext}}{R_{ext} + R_{TH}} \times \frac{n}{n+1} V_{out} = \frac{R_{ext}}{R_{ext} + R_{TH}} V_{feedback,cal}$$

It is seen that the actual feedback voltage is dependent on the external resistance value. The choice of the resistance value should be done keeping this condition in mind.

The actual feedback voltage is compared to the reference voltage for control action. So the error is given by-

$$V_{feedback,cal} = (1 + \frac{R_{Th}}{R_{ext}}) V_{feedback,act}$$

$$Error = \frac{V_{feedback,cal} - V_{feedback,act}}{V_{feedback,act}} = \frac{R_{Th}}{R_{ext}}$$

To minimize the error, the Thevenin's resistance should be as small as possible and the external resistance should be as large as possible. Usually we do not have any control on the value of the external resistance as it depends on the electronic circuit or system to which the feedback signal is connected. On the other hand, the Thevenin's resistance should not be too small as the current through the potential divider and consequently the power loss in it is completely wasted. The combined resistance (R_1 + R_2) should be significantly large compared to the equivalent load resistance of the circuit.

4.6 Interfacing a Digital Signal

The digital signals, whether input or output, are connected to the port pins of the microcontroller. Though the microcontroller operates at 3.3V, the port pins are 5V tolerant. This means that input digital signals up to 5V can be directly connected to the port pin without damaging the port. The SiLab C8051F020 has eight general purpose I/O ports, so a total of 64 pins are available for interfacing digital signals. The 32 port pins of the lower four ports, P0, P1, P2 and P3, have dual functions and can be defined as General-Purpose I/O (GPIO) pins or connections to internal resources such as UART, SPI etc. The programmer needs to define the function of the particular port pin for which it is used. This is achieved through the use of a Priority Crossbar Decoder and the steps to configure it are described below.

Step 1: Allocation of the Pins

The first step is to allocate the port pin for the intended function. This is achieved by using the three crossbar registers, XBR0, XBR1 and XBR2. By setting the bits of the crossbar registers it is possible to assign the port pins to a peripheral. Each crossbar register is an 8-bit register and each bit can be used to allocate pins for different functions such as UARTs, SMbus, Timers etc. in a priority order. The port pins are used in order of priority starting from P0.0 to P3.7. The UART0 has the highest priority and the CNVSTR has the lowest priority as shown in Priority Crossbar Decode table in figure 4.9.

For example, if XBR0.2, which is the UART0EN bit, is set to logic 1, the TX0 and RX0 will be mapped to P0.0 and P0.1 respectively. So by setting XBR0.2 to 1, P0.0 and P0.1 are allocated for the UART0 and cannot be used for any other purpose. In this case it is not possible to assign only TX0 or RX0; they are always allocated as a group. Similarly if UART1 is also used, XBR2.2, which is the UART1EN pin, is to be set to logic 1. If any other peripheral such as SCK, MISO and so on are not used, the TX1 and RX1 will be mapped to P0.2 and P0.3 respectively.

In the Priority Crossbar Decode the priorities are shown from top to bottom in the rightmost column, starting from UART0 being the highest priority to CNVSTR being the lowest priority. The priority for the pins is from P0.0 to P3.7

as shown in the topmost row. Once the crossbar registers are defined to allocate pins for an intended function, the crossbar should be enabled. This is done by setting the XBARE (pin XBR2.6) to logic 1. The port pins which are not allocated are available to use as general purpose input output pin (GPIO).

Step 2: Configuring the Port Pin as Digital Input

In some situations, with the assigned function of the port pin, the pin automatically functions as an input pin. For example, while UART0 is assigned using XBR0.2, the pins P0.0 and P0.1 are reserved for TX0 and RX0. The port pin P0.1 is automatically set as input pin for RX0. For configuring a port pin as a digital input, the following steps are required – set the output mode of the pin as "open-drain" and then write a logic "1" to the associated bit of the port data register. For example if P0.2 is used as a digital input, we have to do the following code:

	P0								P1								P2								P3								Crossbar Register Bits
PIN I/O	0	1	2	3	4	5	6	7	0	1	2	3	4	5	6	7	0	1	2	3	4	5	6	7	0	1	2	3	4	5	6	7	
TX0	●																																UART0EN: XBR0.2
RX0		●																															
SCK	●		●																														SPI0EN: XBR0.1
MISO		●		●																													
MOSI			●		●																												
NSS				●		●																											
SDA	●		●		●		●																										SMB0EN: XBR0.0
SCL		●		●		●		●																									
TX1	●		●		●		●		●																								UART1EN: XBR2.2
RX1		●		●		●		●	●	●																							
CEX0	●		●		●		●		●	●	●																						PCA0ME: XBR0.[5:3]
CEX1		●		●		●		●	●	●	●	●																					
CEX2			●		●		●		●	●	●	●	●																				
CEX3				●		●		●	●	●	●	●	●	●																			
CEX4					●		●		●	●	●	●	●	●	●																		
ECI	●	●	●	●	●	●	●	●	●	●	●	●	●	●	●	●																	ECI0E: XBR0.6
CP0	●	●	●	●	●	●	●	●	●	●	●	●	●	●	●	●	●																CP0E: XBR0.7
CP1	●	●	●	●	●	●	●	●	●	●	●	●	●	●	●	●	●	●															CP1E: XBR1.0
T0	●	●	●	●	●	●	●	●	●	●	●	●	●	●	●	●	●	●	●														T0E: XBR1.1
/INT0	●	●	●	●	●	●	●	●	●	●	●	●	●	●	●	●	●	●	●	●													INT0E: XBR1.2
T1	●	●	●	●	●	●	●	●	●	●	●	●	●	●	●	●	●	●	●	●	●												T1E: XBR1.3
/INT1	●	●	●	●	●	●	●	●	●	●	●	●	●	●	●	●	●	●	●	●	●	●											INT1E: XBR1.4
T2	●	●	●	●	●	●	●	●	●	●	●	●	●	●	●	●	●	●	●	●	●	●	●										T2E: XBR1.5
T2EX	●	●	●	●	●	●	●	●	●	●	●	●	●	●	●	●	●	●	●	●	●	●	●	●									T2EXE: XBR1.6
T4	●	●	●	●	●	●	●	●	●	●	●	●	●	●	●	●	●	●	●	●	●	●	●	●	●								T4E: XBR2.3
T4EX	●	●	●	●	●	●	●	●	●	●	●	●	●	●	●	●	●	●	●	●	●	●	●	●	●	●							T4EXE: XBR2.4
/SYSCLK	●	●	●	●	●	●	●	●	●	●	●	●	●	●	●	●	●	●	●	●	●	●	●	●	●	●	●						SYSCKE: XBR1.7
CNVSTR	●	●	●	●	●	●	●	●	●	●	●	●	●	●	●	●	●	●	●	●	●	●	●	●	●	●	●	●					CNVSTE: XBR2.0
						ALE	/RD	/WR	AIN1.0/A8	AIN1.1/A9	AIN1.2/A10	AIN1.3/A11	AIN1.4/A12	AIN1.5/A13	AIN1.6/A14	AIN1.7/A15	A8m/A0	A9m/A1	A10m/A2	A11m/A3	A12m/A4	A13m/A5	A14m/A6	A15m/A7	AD0/D0	AD1/D1	AD2/D2	AD3/D3	AD4/D4	AD5/D5	AD6/D6	AD7/D7	
									AIN1 Inputs/Non-muxed Addr H								Muxed Addr H/Non-muxed Addr L								Muxed Data/Non-muxed Data								

Fig. 4.9 Priority Crossbar Decode table

```
P0MDOUT.2 = 0; // open drain
P0.2=1;        // write Logic 1
```

Step 3: Configuring the Port Pin as Digital Output

The default state of the output configuration of the port pin is open-drain. The port pin can be configured either as open-drain or as a push-pull. In many situations the push-pull output is required if the port output is required to drive some load. By writing logic "0" or "1" in the associated bit of the port data register, the port pin can be configured as "open-drain" or "push-pull" mode respectively. For example if the port pin P0.4 is set as push-pull output, the following command will do-

```
P0MDOUT.4 = 1;
```

In the following section a few examples are discussed to explain the whole process.

Example 1

In one application, Port 0.0 is used as a digital input and Port0.1 is used as digital output in the push-pull mode. The following few lines of code achieve it.

```
P0MDOUT = 0x00;    // Output configuration for P0, all in open drain
P0 |= 0x01;        // Pin P0.0 is for input so write a '1' to it
P0MDOUT |= 0x02; // Output configuration for P0.1, in push-pull mode
```

Example 2

In one application the UART0 is used and the /INT0 is used for interfacing a digital input. The port pin P0.0 and P0.1 are mapped for TX0 and RX0. The port pin P0.2 is used for /INT0 as digital input. The following few lines of codes are required to achieve it.

```
XBR0 = 0x04;    // Enable UART0 (Pin P0.0 and P0.1)
XBR1 = 0x04;    // Enable /INT0 (Pin P0.2)
XBR2 = 0x40;    // Enable the crossbar

P0MDOUT = 0x00; // Output configuration for P0
P0 |= 0x04;     // Pin P0.2 is for input so write a '1' to it.
```

Example 3

In one application the UART0 is used and both the interrupts /INT0 and /INT1are used for interfacing digital inputs. One digital output in the push-pull configuration is required. The port pin P0.0 and P0.1 are mapped for TX0 and RX0. The port pins P0.2 and P0.3 are used for /INT0 and /INT1 as digital inputs. The port pin P0.4 is used as digital output in push-pull mode. The following few lines of code are required to achieve it.

```
XBR0 = 0x04;    // Enable UART0 (Pin P0.0 and P0.1)
XBR1 = 0x14;    // Enable both /INT0 (XBR1.2) & /INT1 (XBR1.4)
XBR2 = 0x40;    // Enable the crossbar

P0MDOUT = 0x00;// Output configuration for P0
P0 |= 0x0C;     // Pins P0.2 and P0.3 are for inputs so write
                // a '1' to it
P0MDOUT |= 0x10;        // Output configuration for P0.4, in
                        // push-pull mode
```

4.7 Interfacing an Analog Signal

In this section interfacing of an analog signal to the microcontroller has been described. This experiment will program the microcontroller to measure an input analog signal at ADC0 and output it at DAC1.

Using the function generator, an analog signal was fed into the microcontroller at pin AIN0.1 of ADC0. The two signals, input ADC0 and output DAC1, were then measured and compared using the oscilloscope.

Function Generator:	100 Hz sine wave 2 V_{PP} input signal 1 V DC offset
Microcontroller:	8V power supply
Oscilloscope:	Triggering at rising edge Channel 1 = input signal (ADC0) Channel 2 = output signal (DAC1)

ADC0 conversions can be started in four different ways:
1. Software command (writing 1 to AD0BUSY)
2. Overflow of Timer 2
3. Overflow of Timer 3
4. External signal input (rising edge of CNVSTR)

The experiment has been conducted to test out the first three methods, in order to determine which method is more practical to use. The fourth method, using external signal input (rising edge of CNVSTR), is usually not used as it has the lowest interrupt priority.

The *main()* function: The key points to observe in *main()* are calling the various initialization routines and activating the interrupts.

```
void main(void)
{
      EA = 0;                    // disable global interrupts
      Init();                    // general initialization
      Init_Timer3(SYSCLK/50000);     // init Timer 3 to generate
                                     // interrupts
      Init_ADC0();                   // init ADC0
      EA = 1;                        // enable global interrupts

      DAC1CN = 0x80;                 // DAC1 enabled

      while (1)
      {
            DAC1 = adc_result;
      }
}
```

Note: The *main()* will be different for each of the three methods in starting the ADC0 conversion.

Watchdog Timer (WDT)

The WDT are disabled by writing 0xDE followed by, within 4 clock cycles, 0xAD to the WDTCN register. Interrupts were disabled during this procedure to avoid any inadvertent delay between the two writes.

```
WDTCN = 0xDE;         // Disable watchdog timer
WDTCN = 0xAD;
```

External Oscillator Control Register (OSCXCN)

The external crystal oscillator, operating at a frequency of 22.1148MHz, has been set. The external crystal oscillator is enabled by setting CLKSL (OSCICN.3) to 1.

```
OSCXCN = 0x67;        // enable external crystal oscillator
                      // at 22.1148MHz
for (n = 0; n != 255; n++);        // wait for osc to start
while ((OSCXCN & 0x80) == 0);      // wait for Xtal to stabilize
OSCICN = 0x88;                // Internal Osc. disabled (IOSCEN = 0)
```

Digital Crossbar

The crossbar is enabled by setting XBARE (XBR2.6) to logic 1.

```
XBR2 = 0x40; // Enable Crossbar and weak pull-ups (globally)
```

Voltage Reference

For this experiment, the on-chip voltage reference of 2.4V was used. The Reference Control Register, REF0CN, enables or disables the internal reference generator and selects the reference inputs for ADC0.

```
REF0CN = 0x03;        // internal reference buffer on, internal bias
                      // generator on
```

Methods to start ADC0 Conversions:

1) Software command (writing 1 to AD0BUSY)

The AD0BUSY bit is set to 1 to start conversion. It remains set while the conversion is in progress and is restored to 0 when the conversion is completed. The falling edge of AD0BUSY sets the AD0INT interrupt flag and triggers an interrupt (if enabled).

```
ADC0CN = 0x80;        // ADC0 enabled; continuous tracking mode;
                      // ADC0 conversions are initiated
                      // on every write of '1' to AD0BUSY;
                      // ADC0 data is right-justified
AD0BUSY = 1;          // setting AD0BUSY to 1; start conversion
```

In polling method, the AD0INT bit (ADC0CN.5) is polled to determine if a conversion has been completed and then the data is read from the ADC0 register.

```
while ( (ADC0CN &= 0x20) == 0);     // Poll for AD0INT-->1
adc_result = ADC0;                  // read ADC value
```

In an interrupt method, the ADC interrupt service routine is executed when the conversion is complete. Within the ISR, the AD0INT is reset and then the data is read from the ADC0 register.

```
void ADC0_ISR (void) interrupt 15
{
      AD0INT = 0;     // clear ADC conversion complete indicator
      adc_result = ADC0;     // read ADC value

}
```

2) Overflow of Timer 2

Timer 2 is configured in Mode 1 (16-bit auto-reload) to generate an interrupt at regular interval. The count register reload occurs on an FFFFH to 0000H transition and sets the TF2 timer overflow flag. On overflow, the 16-bit value held in the two capture registers is automatically loaded into the count registers and the timer is restarted.

```
void Init_Timer2 (unsigned int counts)
{
      CKCON |= 0x20;              // Timer 2 uses the system clock
      T2CON = 0x00;               // T2CON.0 = 0 Allow Auto-reload on
                                  // Timer2, overflow (CP/RL2)
      RCAP2 = -counts;            // Init reload values in the
                                  // Capture registers
      T2 = 0xFFFF;                // count register set to reload
                                  // immediately at first clock occurs
      IE |= 0x20;                 // IE.5,Enable Timer 2 interrupts (ET2)
      T2CON |= 0x04;              // start Timer2 by setting TR2
                                  // (T2CON.2) to 1
}
```

Setting TR2 to 1 enables and starts the Timer 2. The timer uses the system clock as the clock source. To allow capturing fast changing signals, the system clock is used directly, rather than dividing it by 12. As soon as Timer 2 overflows, the corresponding ISR is executed and the ADC conversion starts. Within the Timer 2 ISR, the TF2 (Timer 2 Overflow flag) is reset.

```
void Timer2_ISR(void) interrupt 5
{
      T2CON &= ~(0x80);       // clear TF2
}
```

The ADC0 is configured to start on Timer 2 overflow as follows-

```
ADC0CN = 0x8C;        // ADC0 enabled; continuous tracking mode;
                      // ADC0 conversions are initiated
                      // on overflow of Timer2;
                      // ADC0 data is right-justified
```

3) Overflow of Timer 3

Timer 3 has only one operation mode, which is in 16-bit auto-reload mode. The operation is essentially the same as for Timer 2, except for slight differences in register names and clock sources.

```
void Init_Timer3(int counts)
{
     TMR3CN = 0x02;          // Stop Timer3; Clear TF3,
                             // use SYSCLK as timebase
     TMR3RL = -counts;       // Init reload values
     TMR3 = 0xffff;          // set to reload immediately
     EIE2 |= 0x01;           // enable Timer3 interrupts
     TMR3CN |= 0x04;         // start Timer3
}
```

Setting TR3 (TMR3CN.2) to 1 enables and starts the Timer 3. The timer uses the system clock as the clock source. As soon as Timer 3 overflows, the corresponding ISR is executed and the ADC conversion starts. Within the Timer 3 ISR, the TF3 (Timer 3 Overflow flag) is reset.

```
void Timer3_ISR(void) interrupt 14
{
     TMR3CN &= ~(0x80);      // clear TF3
}
```

The ADC0 is configured to start on Timer 3 overflow as follows-

```
ADC0CN = 0x84;        // ADC0 enabled; continuous tracking mode;
                      // ADC0 conversions are initiated
                      // on overflow of Timer3;
                      // ADC0 data is right-justified
```

4.8 Discussions

The code was initially written with a *printf* statement to display the ADC0 value. This was to show the effect of the *printf*. The figures 4.10 and 4.11 show the delay due to *printf* command.

Out of the three methods of starting the ADC conversion, AD0BUSY is the best method. Although, all three methods work well, Timer 2 overflow and Timer 3 overflow use the timers which in some applications might be needed for other purposes. In addition, using CNVSTR requires extra setups, such as connecting an external signal to trigger the ADC0 conversion. Whereas, using AD0BUSY is the simplest and most convenient way to start the conversion.

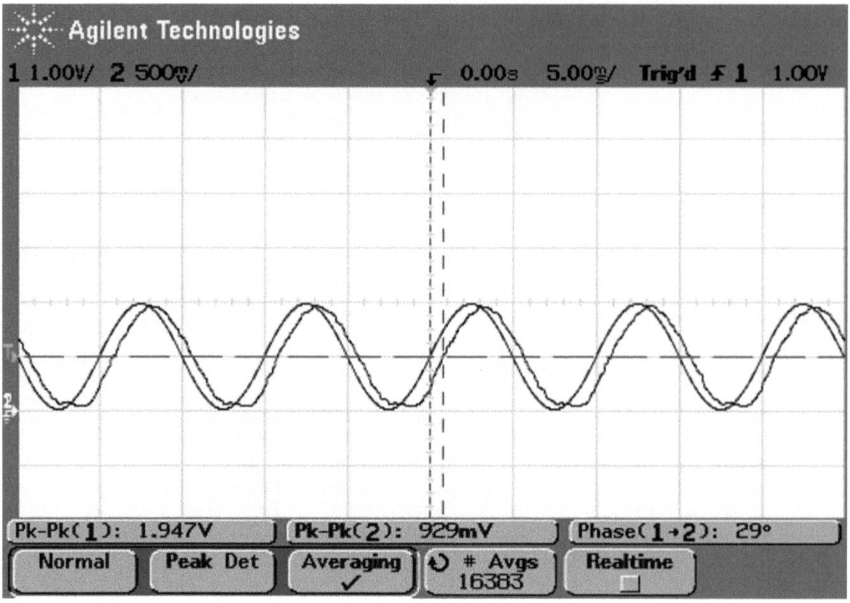

Fig. 4.10 The input and output signals with a *printf* command

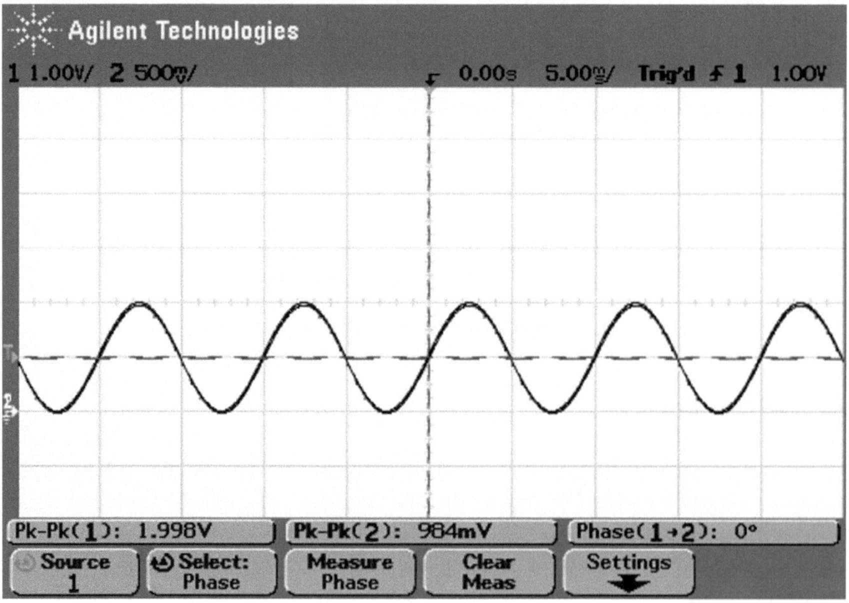

Fig. 4.11 The input and output signals without *printf* command

5

Embedded Microcontroller Based DC Motor Control: A Project Based Approach

5.1 Introduction

This chapter discusses the development of a mixed signal microcontroller based DC motor control project, which has been used for the purpose of teaching motor speed control theory. A problem based learning and teaching approach was taken and our main focus in this chapter is to describe in detail the laboratory part in which the students work in groups on a design project over a complete semester. The major part of the project was the development of software to control the speed of the DC motor. The project also includes the design and development of an over-current protection circuit. If the students complete the project, the end results can show that this type of cooperative problem based teaching and learning can be used to develop their skills in problem solving, enhance team work and their interest in life long learning. The students attain a high level of technical knowledge and become capable of facing the challenges in real life.

Many educators have augmented the conventional textbook presentations by introducing laboratory works in the form of problem based learning to teach the students. In problem based learning a specific problem situation is used to focus the learning activities to achieve the target. Applications have been designed to assist students to teach modern embedded computing subject with the help of computer vision. Computer based simulations and implementations have been developed to illustrate the important practical applications of PID control theory. In addition, some researchers have used fuzzy logic algorithms to teach speed control of DC motor with some success.

The speed control of a DC motor remains a standard component in undergraduate course curriculum in many universities. The project work on DC motor speed control was chosen so that the students get some hands-on experience on measurement and instrumentation (they need to measure the speed of the motor

G. Sen Gupta and S.C. Mukhopadhyay: Embedded Microcontroller Interfacing, LNEE 65, pp. 83–102.
springerlink.com

using tacho-encoder), use of microcontroller as control hardware and implementation of an actual controller to maintain the speed of the motor. In the theoretical part of the course the students are taught the basics of control systems, power electronics, motor fundamentals, advanced electronics and instrumentation. Given the limited resources, the students are divided into groups for the project work. The number of students depends on the resources of the university, but it has been the experience of the authors that each group consisting of 3 students is an optimum number. If more resources are available a minimum number of 2 students in each group may be allowed. It is always better to have a few students in a group as they learn to work in an environment very similar to practical life situations. An overview of the project should be introduced at the beginning of the course and the target for each week should be defined clearly. The students are given access to the laboratory resources so that they can utilize their free off-time to work on the project. The target microcontroller used in this project is Silabs C8051F020. Theory of the microcontroller should be covered in another subject in a previous semester so that the acquired knowledge provides the students with a sound vehicle for gaining an understanding of the project materials.

5.2 Description of the Problem

The main part of the project was to control the speed of a permanent magnet DC motor. The experimental set-up, DC motor, tacho-encoder and the loading mechanism is shown in figure 5.1. The schematic representation is shown in figure 5.2. The controlled voltage is applied across the motor terminals #1 and #2. The speed is measured using a tacho encoder, the operating principle of which is based on Hall effect. A supply source is provided between the +ve and GND terminals and the output (a square wave signal) is available between GND and output terminal. The tacho encoder gives one pulse for every revolution of the motor spindle. This is a real challenge from a control point of view especially at low speeds of operation. The motor is loaded by providing a voltage across the loading terminals. The loading is due to eddy current loss in the aluminum disk which rotates along with the motor and cuts the magnetic flux produced by the electromagnet.

The problem of the speed control of DC motor was chosen because it is physically and mathematically quite straight-forward; the relationships between the applied voltage, speed and the load current are easily understood. The explanation for the need to maintain the speed of the motors is easily perceived by the students. Moreover this is a problem which can be readily implemented with the well-known PID controller. The power circuit used to achieve the objective is shown in figure 5.3. The electronic circuit is a single switch based chopper circuit and the Pulse Width Modulation (PWM) control is used for the control of the speed of the DC motor. The microcontroller board used for this project, along with the expansion board, is shown in figure 5.4.

Fig. 5.1 Experimental set-up – DC motor, tacho encoder and loading mechanism

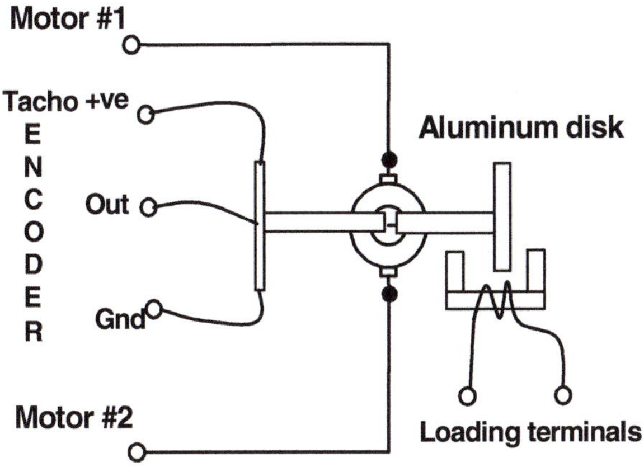

Fig. 5.2 Schematic representation of the set-up

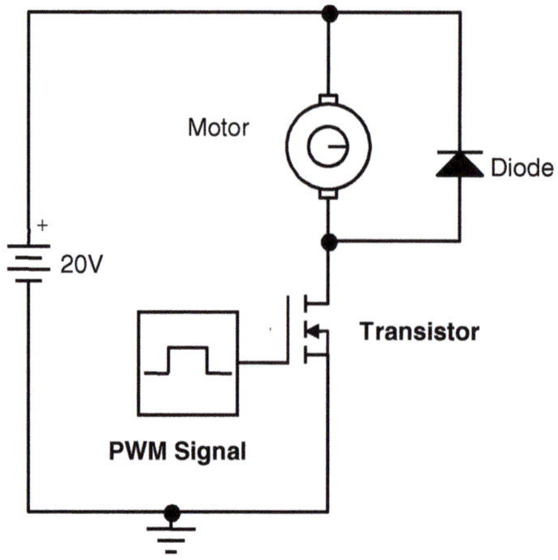

Fig. 5.3 The complete power circuit

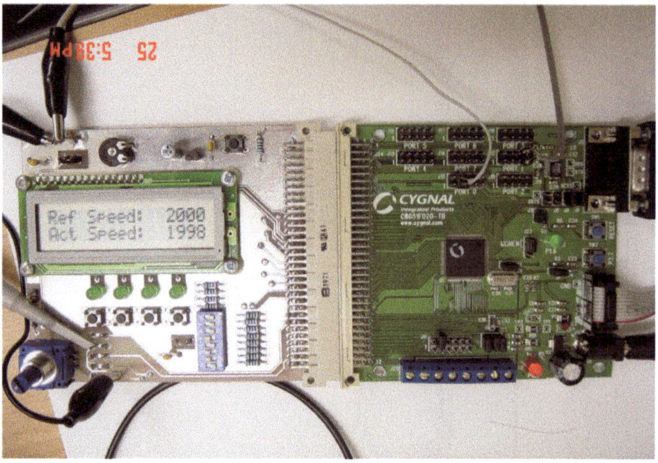

Fig. 5.4 Microcontroller board along with the expansion board

5.3 Motivation of the Project

The motivation of this project came from the philosophy of the engineering curriculum design of Massey University. Most of the courses incorporate project and/or laboratory activities as the curriculum has a stronger emphasis on engineering design practice. This approach helps the students to express their thoughts and try out theirs ideas; they learn the subject matter more deeply, they retain more information of the subject and more importantly their interpersonal skills grow.

5.4 Basic Theory of the Project

The basics of the project work should be taught to the students in one or two hour long lectures. That should consist of the following information:

5.4.1 Speed Control Using Pulse Width Modulation (PWM)

- Used for ***efficient*** DC motor speed control.
- A PWM circuit works by generating a square wave with a variable on-to-off ratio as shown in figure 5.5.
- The average 'on' time may be varied from zero to 100%, but usually doesn't go to 100%. A 100% 'on' time is equivalent to a SHORT circuit of the switch.
- A variable amount of power is transferred to the load (motor), depending on the operation of the motor loading.
- The high value is held during a variable pulse width ***t*** over the fixed period ***T*** where, frequency of the PWM signal = **1/*T***.
- The resulting waveform has a duty ratio, defined as the ratio between the ON time and the period of the waveform, usually specified as a percentage.
- Duty ratio = On-time/ Time period.
- In PWM control, the variable voltage across the armature is applied by switching the transistor (or electronic switch) with different pulse width (ON-time), with fixed period and amplitude (peak value).

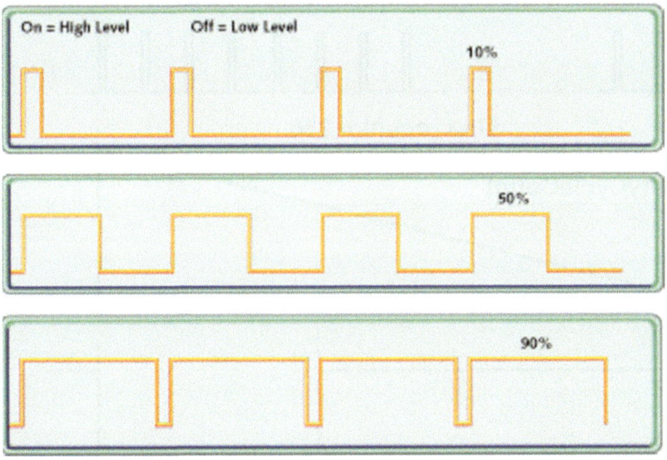

Fig. 5.5 PWM signal of varying duty ratio

- However, due to the motor inductance and resistance, the resulting current through the motor has a small fluctuation around an average value which is dictated by the amount of loading.

- As the duty ratio gets larger, the average voltage gets larger and the motor speed increases.

5.4.2 Generating PWM Signal

The normal method of generating a PWM signal is to compare a saw-tooth wave of the desired frequency with a control signal and is achieved by using op-amp circuits. The generation of a PWM signal using an embedded controller is explained below (see figure 5.6):

- Use Timer 0 in Auto-reload mode so that it overflows at a regular interval and generates an interrupt (lts call it a software ***'tick'***).
- The *PWM_Counter* is incremented in the Timer 0 Interrupt Service Routine (ISR).
- When the *PWM_Counter* value exceeds *dutyCycleCount*, the *PWM_Output* is reset to 0. This is shown in figure 5.6.
- When the *PWM_Counter* value exceeds *MAX_count*, the *PWM_Output* is set to 1.
- 0 <= *dutyCycleCount* <= 256
- Resolution of the duty cycle is 1/256 (approximately 0.39%).
- PWM output is at one of the digital output port pin, example P0.4. The pin 4 of port 0 must be defined as digital output using the Crossbar.
- The PWM output pin must be configured in push-pull mode.

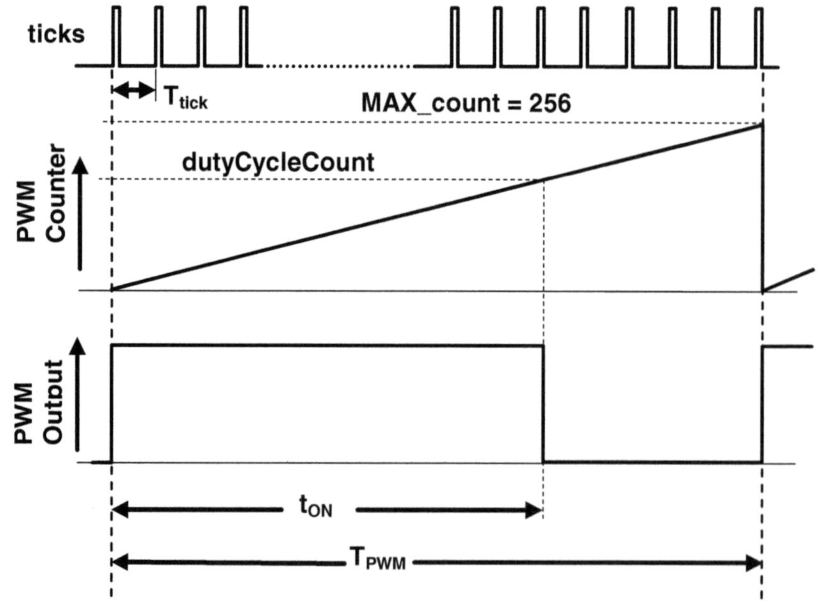

Fig. 5.6 Generation of PWM

The necessary software code for the Timer_0 interrupt service routine is shown below.

```
void Timer0_ISR (void) interrupt 1
{
     //-- clear TF0
     TF0 = 0;

     PWM_counter++;
     if (PWM_counter >= dutyCycleCount)
           PWM_output = 0;

     if (PWM_counter >= MAX_Count)
     {
           PWM_output = 1;
           PWM_counter = 0;
     }
}
```

5.4.3 PWM Frequency: Timer 0 Reload Value

The re-load value for the timer should be correctly calculated.

- For a desired PWM frequency, what should be the 'tick time' (T_{tick})?

$$T_{PWM} = T_{tick} \times 256$$

- If $T_{tick} = 10\ \mu s$, then,

$$T_{PWM} = 10 \times 256\ \mu s$$

$$f_{PWM} = \frac{1}{T_{PWM}} = \frac{1}{10 \times 256 \times 10^{-6}} \simeq 390\text{ Hz}$$

- To produce an interrupt at every 10 μs, what should be the reload value of Timer 0?
- Using a System Clock of 22.1184 MHz,

$$T_{sysclk} = \frac{1}{22118400Hz} \simeq 0.04521\ \mu s$$

- So, the number of clock pulses required is:

$$\#\text{Sysclk pulses} = \frac{10}{0.04521} \simeq 221$$

- Timer 0 reload value = 255 – 221 = 34

5.4.4 Varying the PWM Duty Ratio

While the control of the speed of the DC motor is tested in open loop condition, the potentiometer on the expansion board can be used to set the duty ratio.

- The PWM ON time is changed by changing the value of *dutyCycleCount*, which must be between 0 and 255.
- Use the potentiometer on the expansion board which is connected to the Analog Input (AIN0.2) of ADC0.
- The ADC0 output is a 12-bit data (0 to 4095).
- In the program, read the ADC0 output and divide by 16.

The pictorial representation of the whole process is shown in figure 5.7.

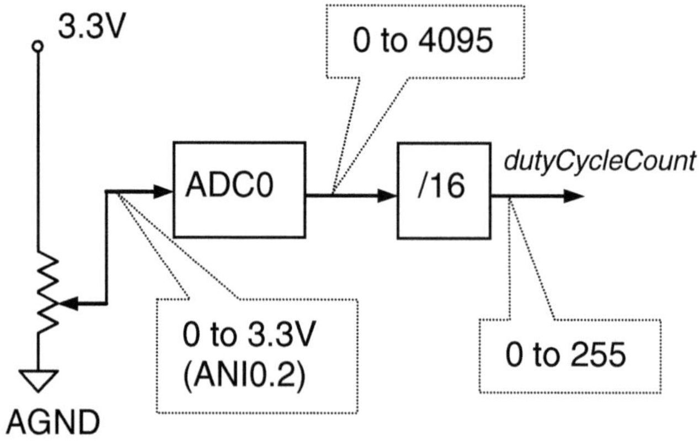

Fig. 5.7 Generation of PWM duty ratio

- There are many ways to initiate the ADC0 conversion:
  - Software command (Writing 1 to AD0BUSY)
  - Overflow of Timer 2
  - Overflow of Timer 3
  - External signal input (rising edge of CNVSTR).

5.4.5 Measuring Motor Speed and Closed Loop Control

- For motor speed control, a *reference speed* is set (usually in rpm)
- The digital speed controller (implemented in software) employs control algorithms to maintain the *actual motor speed* as close as possible to the *reference speed.*
- The controller needs to know the *actual motor speed* so that if there is a difference between the *actual speed* and the *reference speed,* it can take suitable corrective action to minimise the difference in speed.

Tacho-Encoder has been used to get information of the speed. Its characteristics are-

- Hall-effect device.
- Supply: +5V DC
- Output is a square pulse of 50% duty cycle.

It gives one pulse per motor revolution. Slower the motor, larger the time period of the Tacho pulse.

5.4.6 Measuring Actual Motor Speed

General principle of measuring actual motor speed-

- For one Tacho pulse (say, between two rising edges) count the number of fast-occurring ticks (overflows) of a programmed timer. This is explained in figure 5.8.

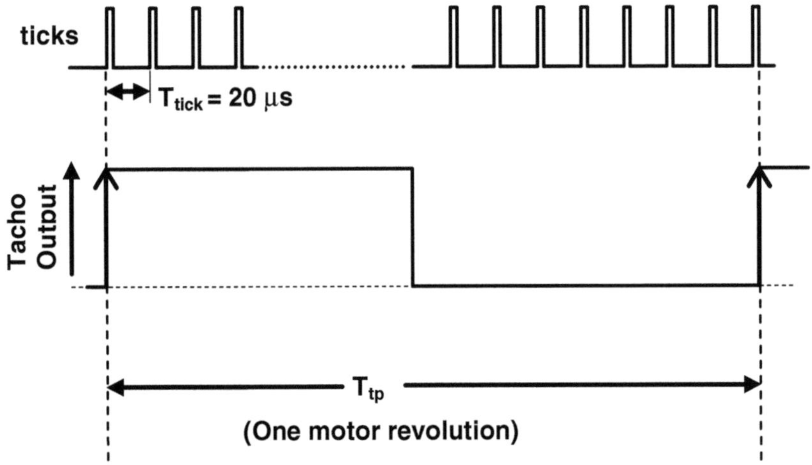

Fig. 5.8 Measurement of actual speed

- Use Timer 1 in auto-reload mode to generate the software interrupts (*ticks*) at 20 μSec interval.
- Calculate the Timer 1 reload value for a System Clock of 22.1184 MHz.
- N: number of ticks counted for one motor revolution.

$$\textit{Actual Motor Speed} \propto \frac{1}{N}$$

$$\textit{Actual Motor Speed} = \frac{K}{N}$$

where,

K is the proportionality constant.

5.4.7 Calculating the Value of K

$$\textit{Actual Motor Speed} = \frac{K}{N}$$

$$\textit{Actual Motor Speed in RPS} = f_{tp} = \frac{1}{T_{tp}}$$

$$\textit{Actual Motor Speed in RPM} = \frac{60}{T_{tp}} = \frac{60}{N * T_{tick}}$$

$$\therefore \quad \frac{K}{N} = \frac{60}{N * T_{tick}}$$

$$\therefore \quad K = \frac{60}{T_{tick}}$$

- If $T_{tick} = 20$ μs, K = 3000000

5.4.8 Counting N (Number of Ticks for One Revolution)

- For every tick, increment a counter (let us call it *TACHO_counter*). This can be done in the ISR of Timer 1.
- Use the Tacho encoder pulse to generate an external hardware interrupt at /INT0.
- Count the number of ticks between two successive Tacho interrupts; this is N.

5.4.9 Setting Motor Reference Speed

- In a digital speed controller, the controller computes the duty cycle of the PWM signal to make the motor run at the *reference speed*.
- We need to input the *reference speed* to the controller.
- This can be done by using the potentiometer on the expansion board which is connected to the Analog Input (AIN0.2) of ADC0.
- The digital output of the ADC0 can be a measure of the *reference speed* in RPM.
- You may want to clamp the *reference speed* to a range of 200 rpm to 2000 rpm.

Figure 5.9 shows the process of generating the *reference speed* for the controller.

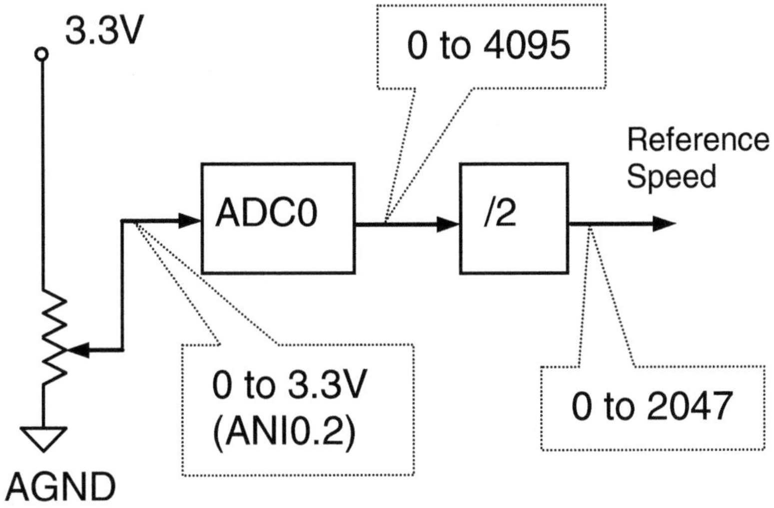

Fig. 5.9 Setting of *reference speed* of the motor

5.4.10 Recording Transient Behavior of Motor

- The digital speed controller calculates the actual speed of the motor which can be displayed on the LCD.
- The actual motor speed is continuously changing, even when the system has stabilized and reached a steady state.
- One set of typical characteristics is shown in figure 5.10.

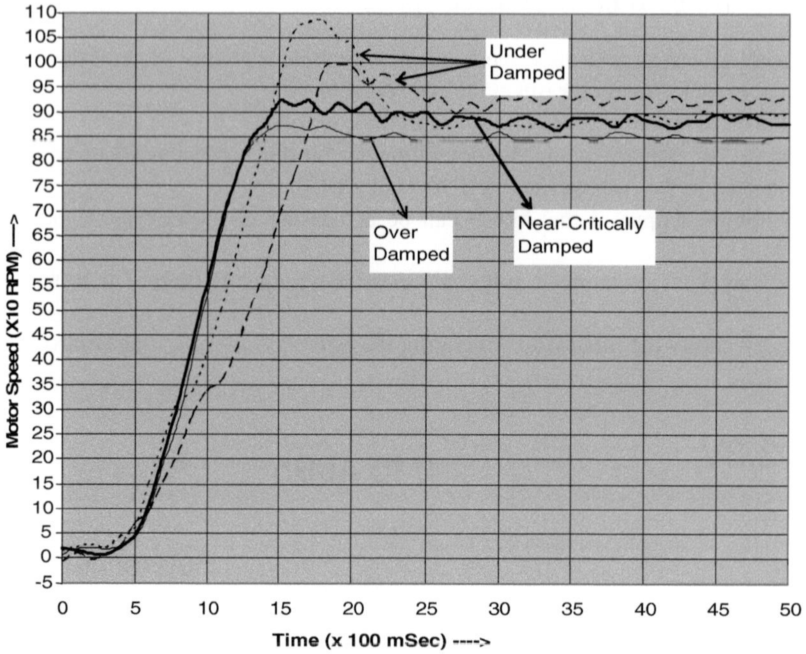

Fig. 5.10 Transient response of the motor

- How are you going to see (and record) the system's transient behaviour?

Display the motor speed on the oscilloscope as an analog voltage.

5.4.11 Displaying Actual Motor Speed as an Analog Voltage on Oscilloscope

- Send the actual motor speed (which is expected to be in the range of 0 to 2000 RPM) to the DAC1.
- The DAC1 analog output will be a measure of the actual motor speed.

Since it is a 12-bit DAC, the input can be in the range of 0 to 4095. Hence the actual motor speed may be multiplied by 2 before presenting at the DAC1 (see figure 5.11). The complete programme structure is shown in figure 5.12.

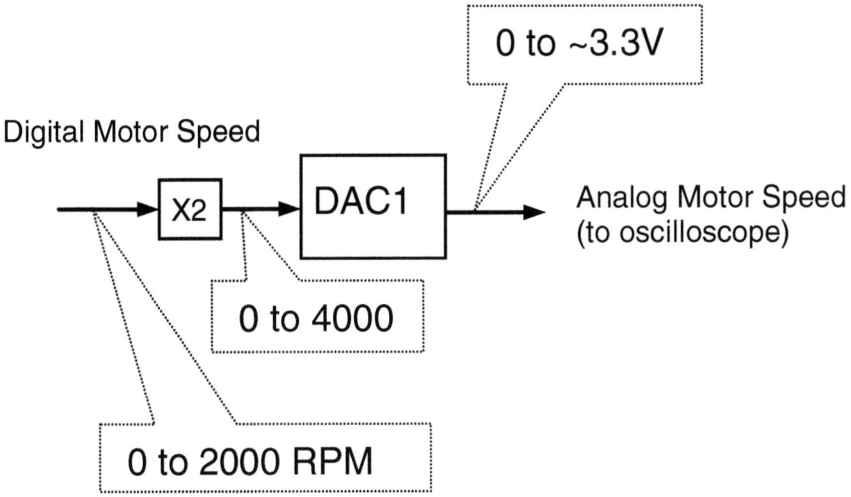

Fig. 5.11 Display of actual motor speed in an oscilloscope

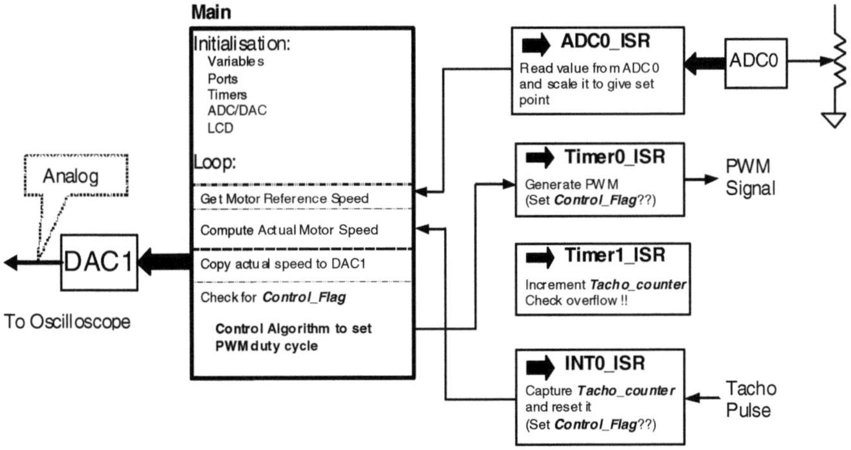

Fig. 5.12 The complete program structure

5.5 Guidelines to the Students

In this section some guidelines are provided so that the students can take the challenge and achieve success. The total activities are summarized as follows-

The speed of the DC motor is controlled or maintained constant by changing the voltage across the armature terminals. In order to do this the actual speed is measured from the signal obtained from a tacho_encoder. The voltage across the motor terminals is varied by changing the ON time of the switch for a fixed time period which is commonly known as PWM (Pulse Width Modulation) control.

For achieving good performance, different types of controllers (P, PI, PID) have been developed and the characteristics of each are studied. Finally the over-current protection has been implemented.

The complete project work has been divided into five activities. To start with, the students are asked to write a program to generate a PWM signal at one of the port pins of the microcontroller. Using DIP toggle switches on the expansion board, they are asked to vary the frequency of the PWM signal and generate

(1) 250 Hz, (2) 400 Hz and (3) 600 Hz.

Using the potentiometer on the expansion board (connected to Analog Input Channel AIN0.2), they are asked to vary the pulse width (duty cycle) from 0% to 100%. Duty Cycle has a step resolution of 1/256. They are asked to display the duty ratio (as a percentage) on the LCD (Liquid Crystal Display).

Some guidance has already been provided to the students for the above activity. 22.1148 MHz external oscillator is used throughout the project. Timer 0 is used in auto-reload mode to generate a software interrupt (let us call it ***PWM_tick***) at every ***X*** micro-seconds. They are asked to calculate ***X***, which will be different for the different PWM frequencies.

$$T_{PWM} = X * 256$$

To generate the ***PWM_tick*** at every ***X*** micro-seconds, they need to calculate the timer reload value.

In the ISR of Timer 0, a counter is incremented (let us call it ***PWM_counter***). When the ***PWM_counter*** value exceeds the threshold value (0 to 255) set by the potentiometer, the PWM output will be reset. The PWM output is set when the ***PWM_counter*** reaches 256.

Once the students are able to achieve the above task they are asked to complete Table 5.1.

Table 5.1 Timer reload value for different frequency settings

PWM Frequency (Hz)	X (μSec)	Timer 0 Reload Value (decimal)
250 Hz		
400 Hz		
600 Hz		

Next they are asked to connect the PWM signal output pin to the gate of the transistor to run the motor. The duty cycle is varied from 0% to 100% and the change of motor speed is observed. They are also asked to note down the motor speed as a function of duty cycle. They repeat this for different PWM frequencies.

- What is the minimum duty cycle at which the motor starts rotating?
- Does this change with PWM frequency?
- Display the 400 Hz PWM signal on the oscilloscope and record it.

By now the students are able to control the motor speed using the potentiometer (to change the PWM duty cycle). They have measured the motor speed from the tacho encoder waveform observed on the oscilloscope. However, to implement a closed-loop motor speed controller, they need to automatically measure the motor speed using the resources of the microcontroller. The measured motor speed (in rpm) needs to be displayed on the LCD.

The following guidance is provided for this task. They are asked to use Timer 1 in auto-reload mode to generate a software interrupt (let us call it ***TACHO_tick***) at every 20 µSec. They need to calculate the timer reload value. For every ***TACHO_tick***, increment a counter (let us call it ***TACHO_counter***). This can be done in the ISR of Timer 1.

Use the tacho encoder pulse to generate an external hardware interrupt. They use external interrupt /INT0.

Count the number of ***TACHO_tick***s between two successive tacho interrupts. This count (let us call it ***N***) is inversely proportional to the motor speed.

Motor Speed = **K/N**
where, **K** is the proportionality constant.

They are asked to calculate the value of K and the Timer 1 reload-value to generate an interrupt every 20 µSec? They calculate and record in Table 5.2.

Table 5.2 Tacho pulses as a function of speed

Motor Speed (RPM)	Tacho Pulse Time Period (µSec)	N
100		
500		
1000		
1500		
2000		

Knowledge of the memory requirements of the various data types is very important. The students are asked to comment on the data type of the variable ***TACHO_counter***.

They can run the motor with the generated PWM of variable duty ratio (using potentiometer) and measure the actual speed of the motor. Display the duty ratio in the 1st row and the measured motor speed in the 2nd row of the LCD.

By now the students have the infrastructure to start implementing a PID controller for motor speed control. Thus far the potentiometer has been used to vary the motor speed by generating a PWM signal of varying duty ratio. In a PID controller, the PWM duty ratio is automatically calculated by the controller to maintain the motor speed at a set reference speed.

The students are asked to modify their program such that the potentiometer output is used to set the reference motor speed in RPM. Clamp the reference speed such that it is between 200rpm and 2000 rpm. They are instructed to use DAC0 to display measured motor speed on the oscilloscope.

They are now capable of implementing a closed loop speed control of the motor. The basic idea is to maintain the speed equal to the reference speed.

Potentiometer sets the *Reference_Speed*

Measure the *Actual_Speed*

Calculate the *Error_Speed* by *Reference_Speed – Actual_Speed.*

Proportional Control (P-Control)

Proportional_Factor = Error_Speed * Kp

where, *Kp* is the gain of the proportional controller.

Duty ratio = Proportional_Factor

Proportional Plus Integral Control (PI-Control)

Integral_Factor = Integral_Factor + Ki * Error_Speed

Now the duty ratio will be decided by

Duty ratio = Proportional_Factor + Integral_Factor

They are now asked to implement a PID control. They need one more variable here to store the previous speed error.

Current_Speed_Error
Previous_Speed_Error

Derivative_Factor = (Current_Speed_Error – Previous_Speed_Error)*Kd

Duty_Ratio = Proportional_Factor + Integral_Factor + Derivative_Factor

At the end of the control cycle they replace the *Previous_Speed_Error* by *Current_Speed_Error*

Previous_Speed_Error = Current_Speed_Error

Note that the D-part doesn't help to reduce the steady state error but helps to improve the transient stability.

They are asked to record data for different values of Kp for P control, for different values of Ki (with a fixed value of Kp) for PI control and different values of Kd (for fixed values of Kp and Ki) for PID control.

Note: The students need to keep in mind that they are using a compiler which may not handle floating point numbers and operations. In that case, the multiplication with a number of less than one should be carefully done using division.

5.6 Outcome of the Project

The project has been a great success and all the groups have successfully completed the project. A few results from their reports are illustrated here. Figure 5.13 shows the speed error as a function of reference speed for different values of proportional gain, Kp. As expected, the magnitude of the error reduces with the increase of Kp.

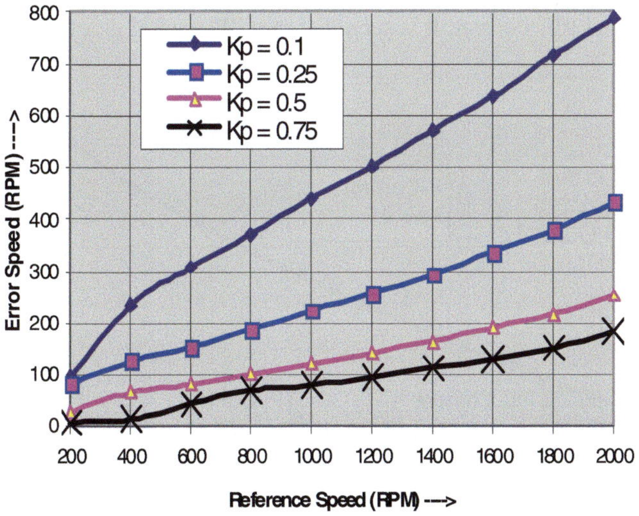

Fig. 5.13 Error speed as a function of reference speed for different values of proportional gains

Figure 5.14 shows the the speed error as a function of reference speed for different values of integral gain, Ki, with a fixed proportional gain Kp of 0.5. It is seen the steady state error is minimum corresponding to Ki of 0.05.

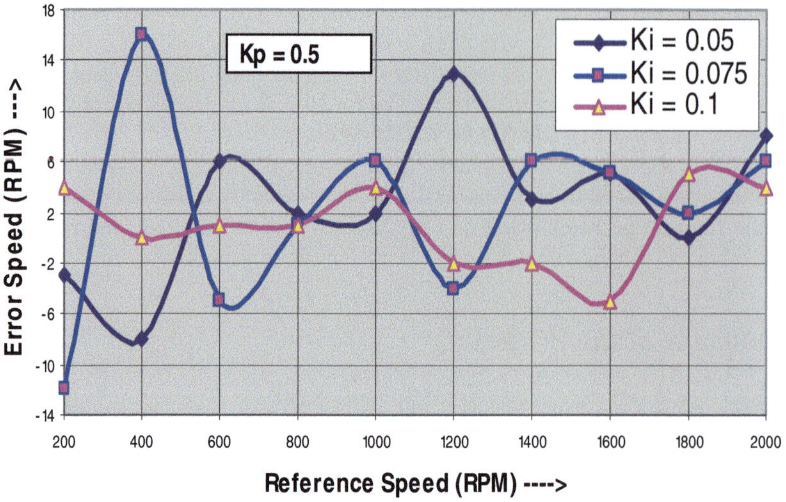

Fig. 5.14 Error speed as a function of reference speed for different Ki values with Kp = 0.5

Figure 5.15 shows the comparative values of speed error as a function of reference speed for P, PI and PID control. It is seen that the PID control gives the best result for the system.

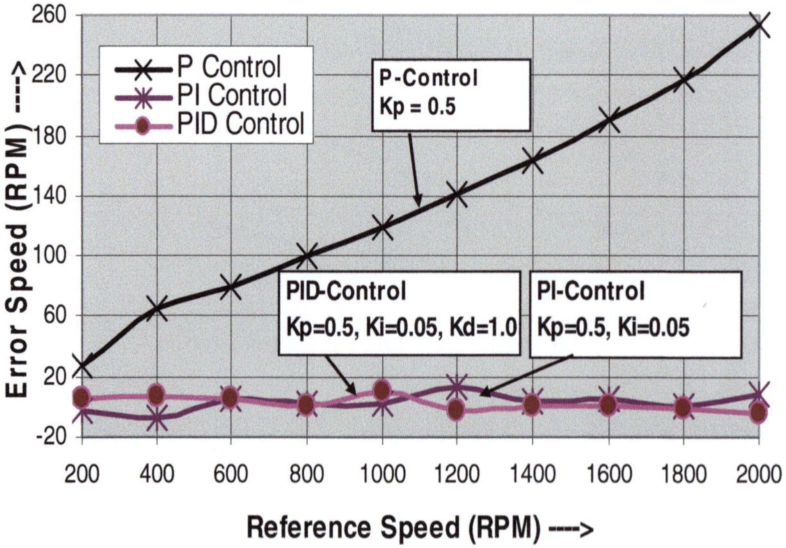

Fig. 5.15 Comparative values of P, PI and PID control

The transient performance of the system was studied with and without the derivative control. The dynamic response of the system improved substantially by introducing the derivative control as can be seen from Figs. 5.16 and 5.17.

In figure 5.16 the peak-to-peak ripple is recorded at 297 mV which translates to a variation in speed of 253 rpm. In Fig. 5.17 the peak-to-peak ripple is recorded at 109 mV which translates to a variation in speed of 93 rpm. The speed variation has dropped by 160 which is an improvement of 63.24%. The steady state speed error was 3 rpm.

Figure 5.18 shows the comparison of transient response for different control algorithms - P, PI and PID.

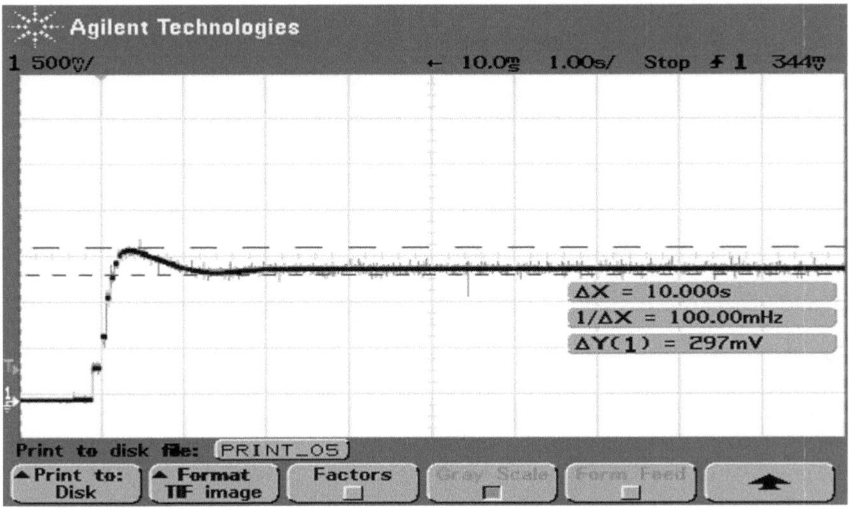

Fig. 5.16 Transient response for PI Control (Kp=0.3, Ki=0.05)

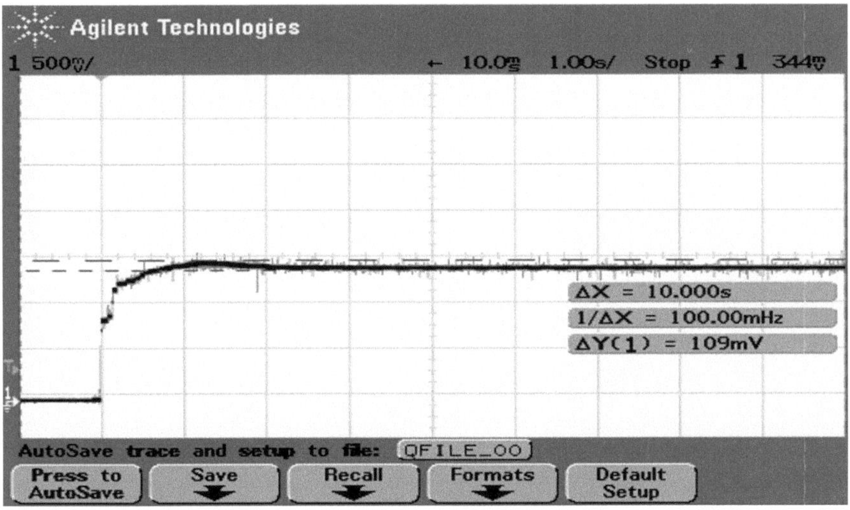

Fig. 5.17 Transient response for PID Control (Kp=0.3, Ki=0.05, Kd=1.0)

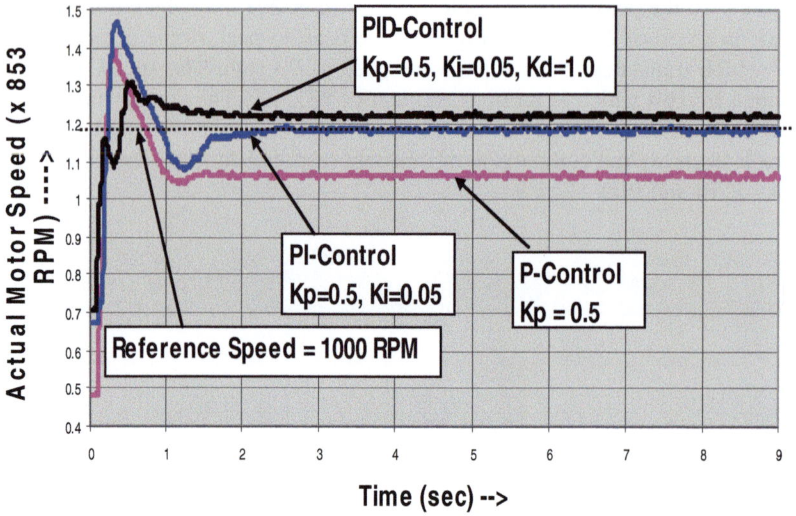

Fig. 5.18 Comparison of transient performance for different control algorithm

6

Embedded Microcontroller Based Switched Mode Power Supply: A Student Project

6.1 Introduction

In undergraduate curriculum of many universities the topic of Switched Mode Power Converter (SMPC) is usually taught as a part of a subject either on power electronics or basic electrical and electronic engineering. At Massey University, New Zealand, the design of a switched mode power supply has been incorporated as a project work in the subject 143.339 (Design for Computer and Communications Systems), a core paper for 3rd year engineering students and the students are required to fabricate the power supply. The subject is offered to students of Information and Telecommunication Engineering, Computer Systems Engineering, Industrial Automation and Mechatronics degree courses at Massey University, New Zealand. The laboratory work contributes 60% of the overall assessment for the course, of which 25% has been allocated for the power supply project. This is a project based paper in which the students are given the problem to solve. Problem (or project) Based Learning (PBL) is the process in which the students focus on the problem and all necessary things are learnt on a need-to-know basis. PBL is ideal for engineering education as it encourages a multi-disciplinary approach to problem solving and develops techniques and confidence in solving problems which is essential for modern engineering practices. Circuit simulation can be used as an aid in teaching the principles of power electronics to improve understanding. But simulation alone will not be sufficient to complete the teaching and there is a need for actual experimentation. The laboratory for conducting real experiment is important. The laboratory component in power electronics can be stimulating and an insightful experience which reinforces classroom learning.

The students design and fabricate a linear power supply which is the basic requirement of the project work. The design, fabrication and implementation of a switched mode power supply itself is a project work on its own merit. Even though an integrated circuit (IC), with a complete switched mode power supply, is now commercially available, from students' learning perspective a lot of things can still be learnt while doing this project.

The control circuit can be implemented using discrete integrated circuits or using a microcontroller. The challenges, design issues and results are presented in

G. Sen Gupta and S.C. Mukhopadhyay: Embedded Microcontroller Interfacing, LNEE 65, pp. 103–126.
springerlink.com

this chapter. In order to make the whole chapter more interesting, the description of the implementation based on discrete circuit is also presented.

The students were divided into groups; each group was asked to design and fabricate a Switched Mode Power Converter, with given specifications, and conduct experiments with it. This was done to develop their design skills in a problem based learning setup. The project was closely linked with practical real-life applications to make it more interesting to the students. The students were allowed the use of design and programming tools such as PROTEL and MATLAB to solve the problem. In order to fully understand the design of the control circuit using microcontroller, the students are asked to first design the controller using discrete integrated circuits. With the design completed using discrete circuits, the students acquire the knowledge to design the controller using a microcontroller. At the end of the project the students were asked to produce a written report.

The students' opinion has been surveyed and it shows the project based approach has enhanced the learning of the subject.

6.2 Description of the Project: Design of Power Supply

The problem given to the students is described in this section.

6.2.1 Specifications of the Problem

Design and fabricate a switched mode power supply with the following specifications:

- Input voltage: 4 V, $\pm$ 20 % (for ex. Solar cell)
- Output voltage: 8 V (for ex. input to microcontroller kit)
- Output current: 100 mA
- Output regulation: $\pm$ 1.0%

6.2.2 Objectives

1. Identify the type of converter and the parameters involved.
2. Design the parameter values.
3. Select the appropriate components.
4. Fabricate the power supply.
5. Design the microcontroller based controller to maintain the output voltage constant irrespective of input voltage and output current changes.

6.2.3 Experiment and Comments

1. Run your power supply with the simulated input (normal power supply) and output (resistive load).
2. Collect data and draw the load regulation characteristics.
3. Collect data and draw the line regulation characteristics.
4. Collect data and draw the efficiency versus output power characteristics.

Comment on the following:

1. Current limiting behaviour of the designed power supply.
2. Thermal shutdown of the designed power supply.

6.2.4 Guidance on the Implementation

The following hints are given to the students on the design and selection of components:

1. Based on voltage and current ratings, choose the diode and MOSFET.
2. Select the capacitor and inductor as close as possible to their designed values.

6.2.5 Experiment with Open-Loop Power Circuit

1. Solder the power circuit on PCB.
2. Connect power from power supply and drive the MOSFET from a Function Generator.
3. Vary the duty ratio of the MOSFET drive from 20% to 80% in steps of 10% with the voltage setting 80%, 90%, 100%, 110% and 120% of the nominal voltage (nominal voltage = 4 V). Connect a load resistance of around 220 Ohms. Note down the output voltage for each setting.
4. Set the input voltage of 4 V and adjust the duty ratio so that you get a nominal output voltage of around 8 V with a load resistance of 220 Ohms. Keeping the duty ratio constant, change the load resistance from 62 Ohms to 1 kΩ (Take 10 readings). Note down the output voltage and input current in each case.
5. From 3, plot the output voltage as a function of duty ratio for different input voltage settings.
6. From 4, plot the output voltage as a function of load current.

6.2.6 Design and Implementation of the Control Circuit

The output voltage will be maintained using a variable duty ratio Pulse Width Modulator (PWM) signal which will be implemented using a microcontroller. The microcontroller is SiLabs C8051F020 operating at a clock frequency of 22.1184 MHz.

6.2.7 Experiment with the Implemented Model

1. Collect data for line regulation and plot it.
2. Collect date for load regulation and plot it.
3. Calculate output power (Vo^2/R), input power ($V_{in}*I_{in}$) and efficiency. Plot the efficiency as a function of output power.

6.2.8 Submission Requirements

The students need to submit the following:

1. The working model.
2. The work-book in which they have designed and collected experimental results.
3. There is a viva-voce (group-basis).

6.3 Design Process

This section describes the implementation of project work with experimental and simulation results. The first thing the students need to do is to select the type of converter. From the specifications it is very clear that they should go for a boost converter to achieve the required 8 V output from the available 4 V input. The power circuit configuration of a boost converter is shown in figure 6.1. The four main elements are the transistor, diode, inductor and capacitor which they need to design and select.

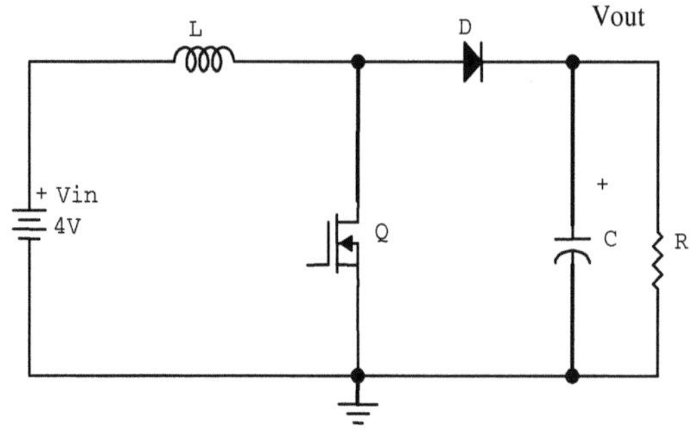

Fig. 6.1 The Boost converter for the project work

Since the voltage and current ratings are not high, they can select high frequency schottky diode and MOSFET of rating 30V and 1A which are usually readily available in the laboratory. The nominal duty ratio is given by-

$$\frac{V_{out}}{V_{in}} = \frac{1}{1-D} \text{ which gives } D = 1 - \frac{V_{in}}{V_{out}} = 0.5$$

The nominal output resistance is given by-

$R = V_{out}/I_{out} = 8/0.1 = 80\ \Omega$

The output voltage ripple for a boost converter is given by-

$$\frac{\Delta V_{out}}{V_{out}} = \frac{DT_s}{RC}$$; T_s is the switching time period.

From the specifications, $\frac{\Delta V_{out}}{V_{out}} \leq 0.05$.

This gives us the value of capacitance C > 1.25 μF. A capacitance of 2.2 μF is chosen.

The inductor current ripple is given by-

$$\frac{\Delta I_L}{I_L} = \frac{D(1-D)^2 RT_s}{L}$$

I_L is the average inductor current. A high value of current ripple will allow choosing a lower value of inductor. Using current ripple of 50%, the calculated inductor value is 200 μH. A standard available inductor of 150 μH is used for the implementation. This will make the current ripple around 66%.

The final circuit configuration used by one group is shown in figure 6.2.

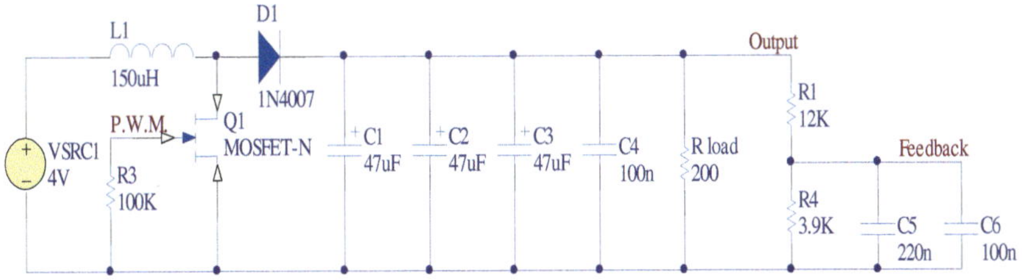

Fig. 6.2 The implemented final circuit diagram

A few capacitors are used in parallel which provides an improved ripple voltage.

The students first run the power circuit under open loop condition before they actually design the controller. Figure 6.3 shows the output voltage as a function of duty ratio while the input voltages are maintained as a percentage of nominal input voltage (five different values - 80%, 90%, 100%, 110% and 120% respectively) under open-loop condition. Figure 6.4 shows the output voltage as a function of load current while the input voltage is kept constant at 4 V and the duty ratio is 50%.

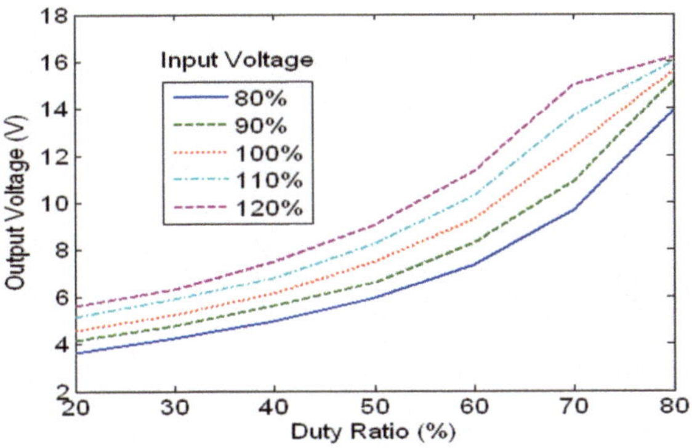

Fig. 6.3 Output voltage as a function of duty ratio for different input voltages

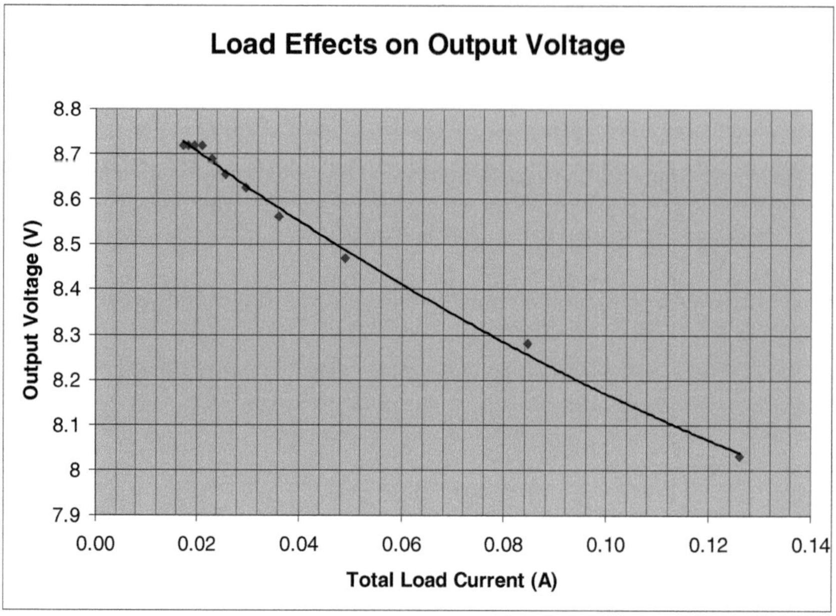

Fig. 6.4 Output voltage as a function of load current under open-loop condition

6.4 Design of a Closed Loop Controller

After working under open loop condition, the students then design the controller. A simplified block diagram describing the whole system is shown in figure 6.5.

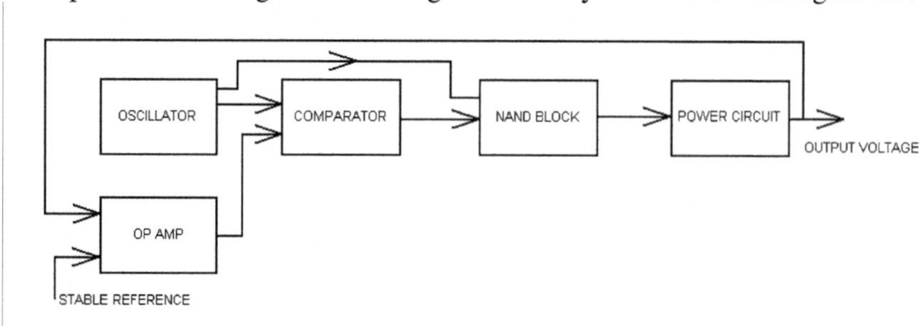

Fig. 6.5 The complete block diagram representation of the switched mode power supply

The functions of the various blocks are described in the following sub-sections.

6.4.1 Oscillator

This is based on a standard Schmitt oscillator with asynchronous switching and is responsible for providing the timing for the control circuit. It serves two functions to this end:

1. Produces a saw tooth waveform that the comparator uses to compare with the error signal generated by the op amp.

2. Produces a 'default' 95% duty ratio that is passed to the NAND block. This is used to provide switching during power-on. When the power is first switched on the error is outside the bounds of the saw-tooth wave and thus the MOSFET is latched on. By providing this 'default' switching cycle we can get the output to increase to a point where the control circuit can begin regulating the output.

6.4.2 Op Amp

This takes a stepped-down measurement of the output and compares it with a stable reference produced by a band gap generator. The error signal produced is used to modulate the gate pulse apparent at the MOSFET via the comparator.

6.4.3 Comparator

This compares the error signal coming from the op amp with the saw tooth waveform produced by the oscillator. Depending on where the DC error signal is in relation to the saw tooth, we can change the duty ratio of the MOSFET.

6.4.4 NAND Block

This block performs two functions:

1. Produces the 'default' PWM for the MOSFET.
2. Decouples the control circuit from the power circuit. We use two NAND gates in parallel to drive the MOSFET. This reduces the time to charge the gate junction and send the device into/out of conduction.

6.4.5 Power Circuit

The circuit shown in figure 6.2 is used as the power circuit. Figure 6.6 shows the fabricated power circuit along with the control circuit.

Fig. 6.6 Fabricated switched mode power supply using discrete components

A few waveforms captured are shown below. Figure 6.7 shows the voltage waveform at the gate of the MOSFET and the ripple voltage at the output. The ripple in the output voltage is negligibly small; it is caused by the switching of the storage components in the boost converter circuit.

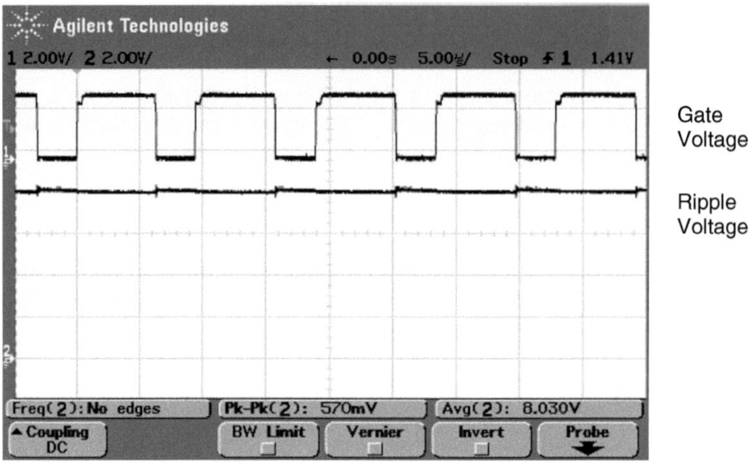

Fig. 6.7 Waveforms at gate and ripple voltage

Figure 6.8 shows the waveforms available at the inputs to the comparator. These two waveforms are used by the comparator to output a pulse width modulated square-wave.

Figures 6.9 and 6.10 show the line regulation and load-regulation characteristics respectively under closed loop control. It is seen that the output is maintained constant within the specified input voltage and load currents.

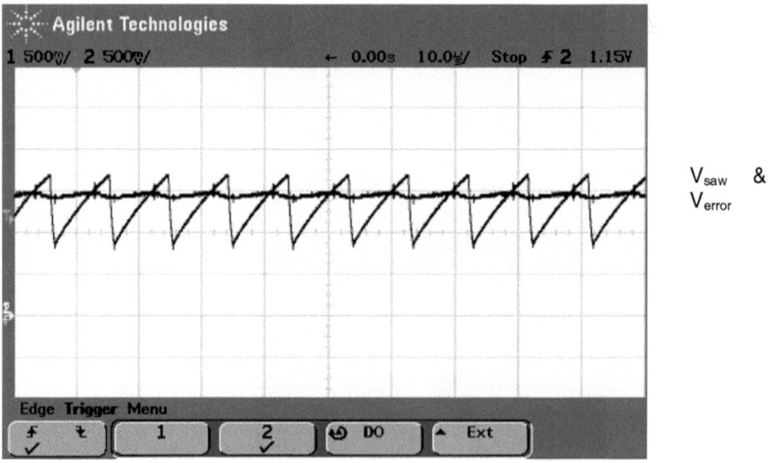

Fig. 6.8 Waveforms at the input of the comparator

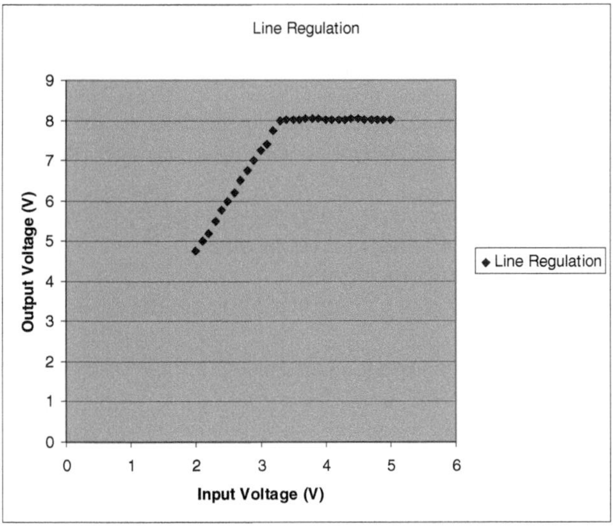

Fig. 6.9 Line regulation characteristics at closed-loop control

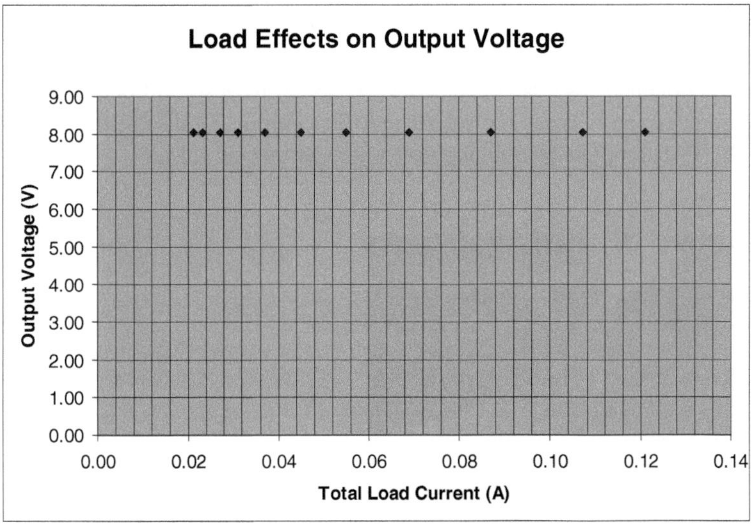

Fig. 6.10 Load regulation characteristics at closed-loop control

6.5 Implementation of an Embedded Microcontroller Based Switched Mode Power Supply

After the fabrication of the switched mode power supply using discrete circuit, the controller can be replaced by the microcontroller. The basic changes that are needed to be made are as follows:

1. the feedback voltage need to be input to the microcontroller,
2. the error voltage is calculated in the microcontroller,
3. The error voltage is used to pass through a controller to obtain the desired duty ratio.
4. The Pulse Width Modulation has been implemented using a timer as discussed in the previous chapter.
5. The current signal has been used after a comparison with a reference voltage to protect the MOSFET.

It is a challenge to implement the pulse width modulation of a switching frequency of 100 kHz with a 22.1184 MHz microcontroller. Of course, a lower switching frequency such as 20 kHz can be chosen for successful implementation.

Figure 6.11 shows the power circuit interfaced with the microcontroller. The output voltage, as well as the output current, is fed as inputs to the microcontroller. The output voltage, as well as the voltage signal corresponding to the output current, are taken as analog signals and connected to the ADC channels. The PWM signal is obtained from port P1.2 which is set up in push-pull output mode. After all the processing is done, the microcontroller outputs a pulse to the gate of the MOSFET. The complete program is given in the appendix.

Fig. 6.11 The fabricated power circuit interfaced with the microcontroller

Figure 6.12 shows the switching ripple at the output voltage. There are some high frequency spikes present due to the switching of the MOSFET. First the feedback was taken from the output without any filter circuit and it was not possible to get a stable operation. The presence of ripple at the output, as is shown in figure 6.13, is considered to be the cause. The filter capacitors of very low cut-off frequency are added at the output as is shown in figure 6.2, to achieve stable operation of the power supply. Figure 6.14 shows the waveform of the feedback voltage after filtering.

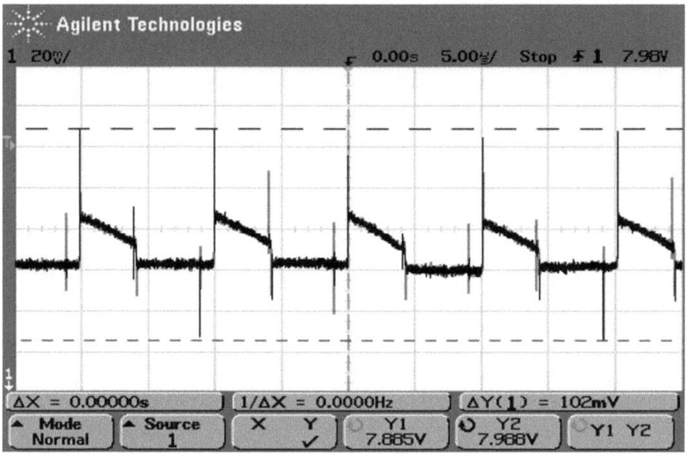

Fig. 6.12 The switching ripple in the output voltage

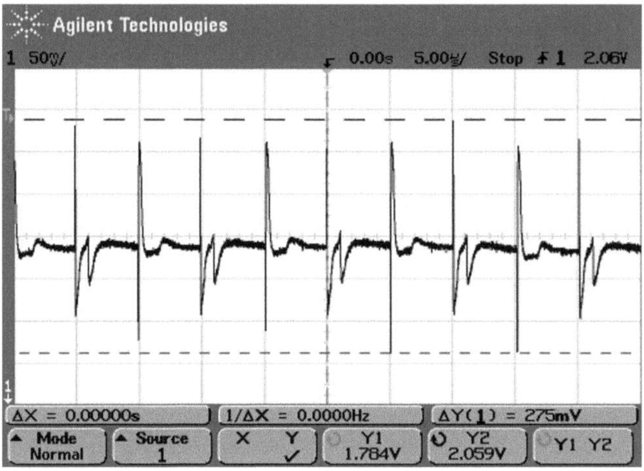

Fig. 6.13 The feedback voltage before filtering

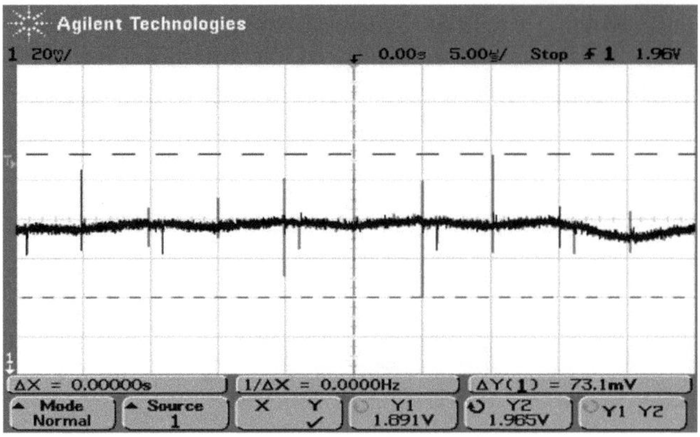

Fig. 6.14 The waveform of feedback voltage after filtering

Figure 6.15 shows the line regulation characteristics under closed-loop condition. The maximum variation of output voltage is within +/- 1% of the nominal output voltage.

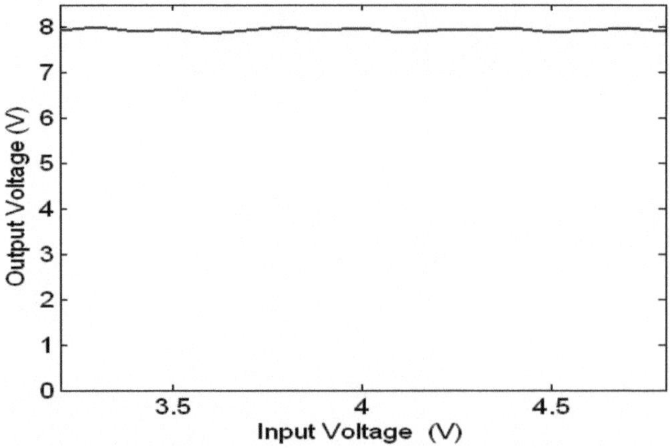

Fig. 6.15 Line regulation characteristics under closed loop condition

Figure 6.16 shows the load regulation characteristics under closed-loop condition. It is seen that the maximum variation of output voltage is within +/- 1% of the nominal output voltage. Figure 6.17 shows the variation of efficiency as a function of power output. The maximum efficiency reaches around 80%. Since the output is less than 1 Watt, this range of efficiency is not totally unexpected.

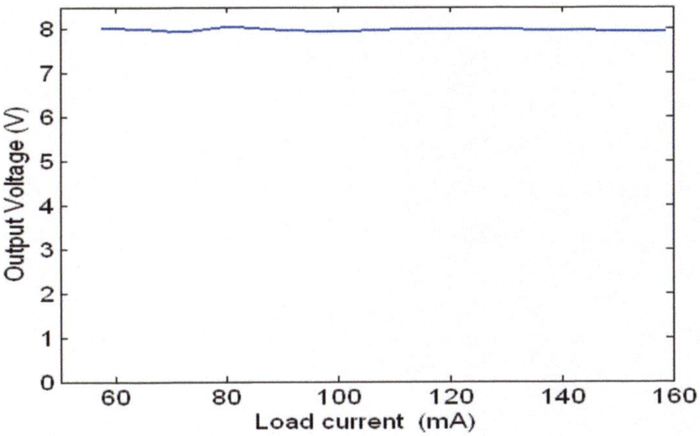

Fig. 6.16 Load regulation characteristics under closed loop condition

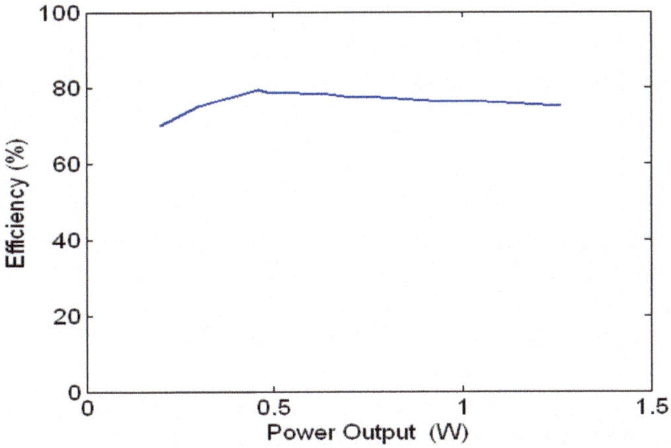

Fig. 6.17 The variation of efficiency as a function of output power

Figure 6.18 shows the picture of a group of students working on the project. The fabricated power circuit and the microcontroller board can also be seen in the picture. The whole project was implemented in three weeks which is the time allocated for it in the teaching schedule.

Fig. 6.18 Students working on project in the laboratory

6.6 Comments

The students were asked to write down the different design issues, challenges and the way they attempted to solve the problems. The following section has been taken from a student report.

6.6.1 Design Issues

There are a few areas of this project that involved making design choices. Firstly, we needed to consider which components to use in building the power supply hardware. On the software side, we had some choices how to generate the PWM control signal, and how to vary its duty ratio. Finally, we needed to think about the interface between our power supply and the microcontroller.

6.6.2 Challenges of the Project Implementation

The following points are considered as challenging for this project:

(i) **100 kHz -** There were a number of approaches that they could have taken to generate a variable duty ratio 100 kHz control signal. We found that only approaches which allowed independence of duty ratio control and signal production would work as duty ratio control took a huge amount of processor time. Ideally these operations would occur concurrently.

(ii) Combining independent power supplies - For most of the project, we did not think about the fact that we were powering the power supply and the microcontroller from two independent power supplies. We did not realize that the independence meant that our power supply voltage could float independently of the microcontroller reference setting.

(iii) **Oscilloscope probe ground leads –** The length of the ground leads on the oscilloscope probes significantly accentuated spikes. In some cases, it made problems appear worse than they actually were.

(iv) **Capacitive coupling** – We encountered capacitive coupling problems in this project when changes in the output of our PWM signal produced spikes in the analogue input. This was tracked back to the signal routing on the microcontroller expansion board.

(v) **ADC multiplexer switching** – We discovered that changing ADC channel produced large spikes on the channels being switched off and on. This must have been the result of capacitive coupling of switching signals and channels in the ADC's analogue multiplexer.

(vi) **Equivalent series resistance** – We learnt about E.S.R. and its effects in the context of selecting our capacitor.

(vii) **Control**

- We used software and hardware filters to remove noise from our feedback; these were necessary for output stability.
- We used an implementation of PWM generation that allowed signal generation to be independent from duty ratio control.
- We used a control system based on the equation: newRatio = oldRatio + k*error. This approach incrementally **adjusts** the duty ratio based on an error.

6.7 Conclusions

In this chapter a student project on design and implementation of a switched mode power supply has been described. First the controller has been implemented by using discrete components and then the power circuit has been interfaced with embedded microcontroller. The implementation of the project using embedded microcontroller was a real challenge but the students have succeeded in implementing it.

A6 Chapter Appendix

Explanation of the implementation of a few parts and the complete listing of the program code is given here.

A6.1 Microcontroller Setup

The microcontroller should be setup to use the external oscillator; this gives the highest frequency to work with. Set OSCXCN to 0x67, this will set the microcontroller to use the external oscillator and set the oscillator to greater then 6.7 MHz.

The next step is to setup the port that will be used. To do this, set up the cross-bar SFR's XBR0, XBR1 and XBR2 to 0x04, 0x80 and 0x40 respectively.

Port 1 can be set for analogue input, by 'ANDing' the port 1 control register with 0xFE. This will set P1.0 to analogue input i.e. P1MDIN &= 0xFE, other then setting P1.0 to analogue input the ports can be set to outputs, i.e. PxMDOUT = 0x00, where x = 0 to 3.

A6.2 Reference Voltage

The reference voltage used is the internal reference voltage of the C8051F020. This can be done by setting the special function register (SFR) REF0CN to 0x07 (Hex). This will cause ADC0 to use the internal bias generator. If the external potentiometer (provided the extension board exists) is used to set up the reference signal, the potentiometer should be configured through the ADC channel.

A6.3 Generation of 100 kHz PWM

To generate the 100 kHz PWM signal, Timer 3 is used in auto-reload mode. The frequency of the PWM is set by the value in TM3RL. To generate 100 kHz TM3RL is set to 0xFF91 (Hex).

A6.4 Feedback Voltage

The feedback voltage that comes from the output of the circuit needs to be divided to bring it within the operating range so that the input to the microcontroller doesn't exceed the allowable limit. A 10k/1k voltage divider was used to ensure that an adequate range of voltages could be measured. Also this resistance is chosen to ensure that the excessive current with respect to the full load current doesn't flow through the feedback resistances.

A6.5 Implementation of PWM

The PWM is implemented by changing the reload value in timer 3 depending on which part of the cycle the PWM is in.

```
void Timer3_ISR (void) interrupt 14
{
        int lwReload;
        lwReload = 221-ADCAdj;
        if ((P1 & 0x02) == 0x02) //if P1 is high go low
        {
                TMR3RL = 65315 + (lwReload) ;
                //low time reload time
                P1 &= 0xFD;
                //set output pin low
                TMR3CN &= ~(0x80);   //-- clear TF3
        }
        else
        {
```

```
13                      TMR3RL = 65315 + (ADCAdj + Variance);
                        //high time reload time
14                      P1 |= 0x02;             //set output high
15                      TMR3CN &= ~(0x80);      //-- clear TF3
16              }
17      }
```

The variable *ADCAdj* is the current value of ADC0, this is a value between 0 – 255. This is the value that the potentiometer is set to.

Line 3&4: the variable *lwReload* is the time that the PWM should stay LOW for. The value of 221 is 65,536 – 65315, thus if the value of *ADCAdj* + *Variance* = 0 then the output should be 0. These values were chosen to keep the frequency of the PWM as close to 100 kHz as possible. The PWM will usually stay within ±2% of 100 kHz.

Line 5&11: test if the output pin is high; if so change the output to LOW, else go HIGH.

Line 7, 8, 9, 13, 14 & 15: set the new timer 3 reload value, set the output pin to the appropriate value and reset the timer 3 interrupt flag.

A6.6 Control Loop

The control loop for this system consists of:

Input Voltage -> ADC0 -> Compare to Goal Voltage -> Generate Variance -> Change PWM -> Input Voltage

A6.7 Listing of the Complete Program Code

```
#include <c8051f020.h>
#include <stdio.h>
//-----------------------------------------------------
// 16-bit SFR Definitions for 'F02x
//-----------------------------------------------------
sfr16 DP = 0x82;                // data pointer
sfr16 TMR3RL = 0x92;            // Timer3 reload value
sfr16 TMR3 = 0x94;              // Timer3 counter
sfr16 ADC0 = 0xbe;              // ADC0 data
sfr16 ADC0GT = 0xc4;            // ADC0 greater than window
sfr16 ADC0LT = 0xc6;            // ADC0 less than window
sfr16 RCAP2 = 0xca;             // Timer2 capture/reload
sfr16 T2 = 0xcc;                // Timer2
sfr16 RCAP4 = 0xe4;             // Timer4 capture/reload
sfr16 T4 = 0xf4;                // Timer4
sfr16 DAC0 = 0xd2;              // DAC0 data
sfr16 DAC1 = 0xd5;              // DAC1 data

//-----------------------------------------------------
// Global DEFINES
//-----------------------------------------------------
#define uchar unsigned char
#define SYSCLK  22118450        // system clk freq in Hz
#define LCD_DAT_PORT  P6        // LCD is in 8 bit mode
#define LCD_CTRL_PORT P7        // 3 control pins on P7
```

```
#define RS_MASK 0x01         // for LCD_CTRL_PORT
#define RW_MASK 0x02
#define E_MASK  0x04

//-----------------------------------------------------
// Global MACROS
//-----------------------------------------------------
#define pulse_E();\
        small_delay(1);\
        LCD_CTRL_PORT = LCD_CTRL_PORT | E_MASK;\
        small_delay(1);\
        LCD_CTRL_PORT = LCD_CTRL_PORT & ~E_MASK;\

//-----------------------------------------------------
// Global Variables and Constants
//-----------------------------------------------------
int i,ADC0_reading, ADC1_reading;
int ADCAdj;
int CurrVoltage, GoalVoltage, Variance;
const int InputVoltage = 4; //set the input voltage as a constant

//-- function prototypes ----------------------------
void Init_Clock(void);        //-- initialise clock to use
                              //   external crystal oscillator
void Init_Port(void);         //-- Configures the Crossbar
                              //    and GPIO ports
void Init_Timer3(void);
void Timer3_ISR(void);        //-- ISR for Timer 3
void init (void);
void Init_ADC0(void);
void Init_ADC1(void);
void ADCtest(void);
void ADC1test(void);

//------ LCD function prototypes
void small_delay    (char d);   // about 0.34us per count @22.1MHz
void large_delay    (char d);   // about 82us   per count @22.1MHz
void huge_delay     (char d);   // about 22ms   per count @22.1MHz

main()
{
        int Count;
        Count=0;
        init();
        for (;;)  //-- go on forever
        {
                //-- test if input is ready
                if ((ADC1CN & 0x1F) == 0)
                {
                        ADC1test();
                }
                //-- test if pot is ready
                if (!AD0BUSY)
                {
                        ADCtest();
                        ADCAdj = ADC0_reading;
                }

                //-- current voltage at the input
```

```
            CurrVoltage = (ADC1_reading - 3) * 11 / 10;
            GoalVoltage = InputVoltage * (1000 / (100 -
             ((ADC0_reading + 15) * 100 / 255))); //Vo=Vi(1/(1-D))

            //-- set limits on the goal voltage
            if (GoalVoltage > 40*InputVoltage)
                    {GoalVoltage = 40*InputVoltage;}
            if (GoalVoltage < 10){GoalVoltage = 40;}
            // max is input voltage * 4, min input voltage + 1volt

            if (((GoalVoltage - CurrVoltage) > 2) || ((GoalVoltage
                                            - CurrVoltage) < -2))
            {
                    Variance = (GoalVoltage - CurrVoltage);
                    //Count++;
            }
            else Variance = 0;

            /*if (Variance > 0 && Count > 10)
                    Variance = Variance + (Count - 9);
            else if (Variance < 0 && Count > 10)
                     Variance = Variance - (Count - 9);
            else Count = 0;*/

            // goal voltage based on the duty ratio of the PWM
            // signal and the input voltage
        }
        return(0);
}
void init (void)
{
        i = 0;
        Init_Clock();
        Init_Port();
        Init_Timer3();

        //-- Initialise ADCs ---------------------------------------
        REF0CN = 0x07; //-- Enable internal bias generator
                       // and internal reference buffer
                       //   Select ADC0 reference from VREF0 pin
        Init_ADC0();
        Init_ADC1();
        //--------------------------------------------------------

        EIE2 = 0x01;   // Turn off all interrupts except
                       // timer 3 interrupt
        EA = 1;        //-- enable global interrupts
        WDTCN = 0x07;  // Watchdog Timer Control Register
        WDTCN = 0xDE;  // Disable watch dog timer
        WDTCN = 0xAD;
        OSCXCN = 0x67; // EXTERNAL Oscillator Control Register
        while ((OSCXCN & 0x80) == 0); // wait for XTAL to stabilize

        OSCICN = 0x0C;   // Internal Oscillator Control Register

        //-- Port 7-4 I/O Lines
        P74OUT = 0x48;  //Output configuration for P4-7
                        //(P7[0:3] Push Pull) - Control Lines for LCD
                        //(P6 Open-Drain)- Data Lines for LCD
```

```
                         // (P5[7:4] Push Pull) - 4 LEDs
                         // (P5[3:0] Open Drain) - 4 Push-Button
                         // Switches (input)
                         // (P4 Open Drain) - 8 DIP Switches (input)
//-- Write a logic 1 to those pins which are to be used for input
      P5 = 0x00; // turn off LEDs
      P4 = 0xFF;
}

//--------------------------------------------------------------------
// delay routines
//--------------------------------------------------------------------
void small_delay(char d)
{
      while (d--);
}

void large_delay(char d)
{
      while (d--) small_delay(255);
}

void huge_delay(char d)
{
      while (d--) large_delay(255);
}
//====================================================================

void Init_Clock(void)
{
//    int count;
      OSCXCN = 0x67; //-- 0110 0111b
                     //-- External Osc Freq Control Bits (XFCN2-0)
                     //   set to 111 because crystal freq > 6.7 MHz
                     //-- Crystal Oscillator Mode (XOSCMD2-0) set
                     //   to 110
      //-- wait till XTLVLD pin is set
      while ( !(OSCXCN & 0x80) );

      OSCICN = 0x88; //-- 1000 1000b
                     //-- Bit 2 : Int Osc. disabled (IOSCEN = 0)
                     //-- Bit 3 : Uses External Oscillator as
                     //   System Clock (CLKSL = 1)
                     //-- Bit 7 : Missing Clock Detector Enabled
                     //  (MSCLKE = 1)
      CKCON = 0x00;
}

void Init_Port(void)  //-- Configures the Crossbar and GPIO ports
{

      // Configure the XBRn Registers
      XBR0 = 0x04;   //-- Enable UART0 (which uses P0.0 and P0.1)
      XBR1 = 0x80;
      XBR2 = 0x40;   // Enable the crossbar, weak pullups enabled

      // Port configuration (1 = Push Pull Output)

      P1MDIN &= 0xFE; //p1.0 is set as analog input
```

```
        P0MDOUT = 0x00;  //-- Enable TX0 as a push-pull output
        P1MDOUT = 0x02;  // Output configuration for P1
                         // (Push-Pull for P1.6)
        P2MDOUT = 0x00;  // Output configuration for P2
        P3MDOUT = 0x00;  // Output configuration for P3
}

//-- Configure Timer3 to auto-reload and generate an interrupt at
//-- interval specified by <counts> using SYSCLK/12 as its time base.
void Init_Timer3 (void)
{
        TMR3CN = 0x02;      //-- timer 3 uses system clock
        TMR3    = 0xFFFF;   //-- set to reload immediately

        TMR3RL = 0xFF91  //-- set the timer 3 reload value, this sets
                         //  the frequency to approx 100 kHz if using
                         // the 22 MHz external oscillator

//      EIE2   &= ~0x01; //-- disable Timer3 interrupts
        EIE2   |= 0x01;  //-- enable Timer3 interrupts
        TMR3CN |= 0x04;  //-- start Timer3 by setting TR3 to 1
}

//-- Interrupt Service Routine
void Timer3_ISR (void) interrupt 14
{
        int lwReload;
        lwReload = 221-ADCAdj;
        if((P1 & 0x02) == 0x02) //if P1 is high go low
        {
                TMR3RL = 65315 + (lwReload) ; //low time reload time
                P1 &= 0xFD;                   //set output pin low
                TMR3CN &= ~(0x80);  //-- clear TF3
        }
        else    //else P2 go high
        {
                //high time reload time
                TMR3RL = 65315 + (ADCAdj + Variance);
                P1 |= 0x02;    //set output high
                TMR3CN &= ~(0x80);  //-- clear TF3
        }
}

void Init_ADC0(void)
{
                        // Internal Temperature Sensor ON
        ADC0CF = 0x80; //-- SAR0 conversion clock=1.3MHz approx.,
                        //   Gain=1
        AMX0CF = 0x00; //-- 8 single-ended inputs (but for temp
                        //  sensing this really doesn't matter)
        AMX0SL = 0x00; //-- Select AIN0.2 (Potentiometer
                        // on the Expansion Board)
        ADC0CN = 0x81; //-- enable ADC0, Continuous Tracking Mode
                        //   Conversion initiated on Timer 3 overflow,
                        //   ADC0 data is left justified
        AD0INT = 0;    //-- clear ADC0 conversion complete interrupt
                        //   flag
        AD0BUSY = 1;
}
```

```
void Init_ADC1(void)
{
        ADC1CF = 0x81; //--SAR1 conversion clock= 941 Khz, Gain =1
        AMX1SL = 0x00; //--Select AIN1.0 input
        ADC1CN = 0x80; // enable ADC1, contiunous tracking mode,
                       // conversion on timer3 overflow
        ADC1CN |= 0x20;//Clear AD1INT
        ADC1CN |= 0x10; // Set busy bit
}

void ADCtest(void)
{
        int lowbit;
        AD0INT = 0;//-- clear ADC0 conversion complete interrupt flag
        ADC0_reading = ADC0H;
        lowbit = ADC0L;
        AD0BUSY = 1;
}

void ADC1test(void)
{
        ADC1CN &= 0xDF;//clear ADC1 conver. complete interrupt flag
        ADC1CN |= 0x10;         // Set busy bit
        ADC1_reading = ADC1;
}
```

A6.8 Working Waveforms

Figure 6.19 shows the PWM that is generated by the microcontroller, as the trace from the scope moves away from the starting point the PWM seems to ‘shimmer’. This is because the PWM is being altered to keep the voltage constant. The dynamic changes in the PWM are relatively small and thus the voltage is kept constant.

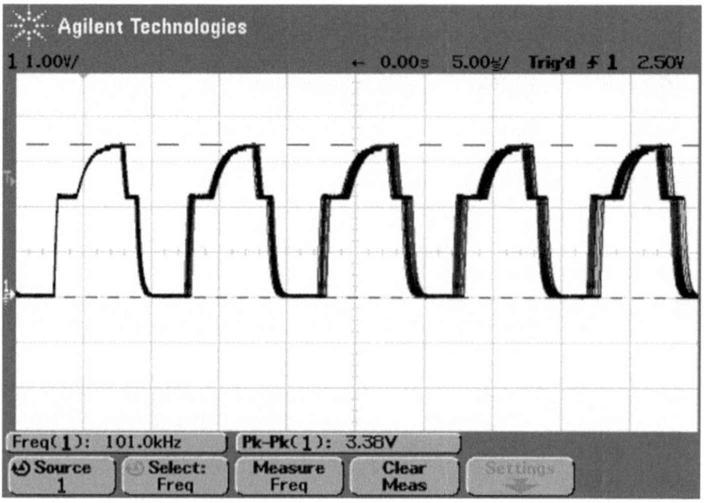

Fig. 6.19 PWM signal, output is 8V with 4V input

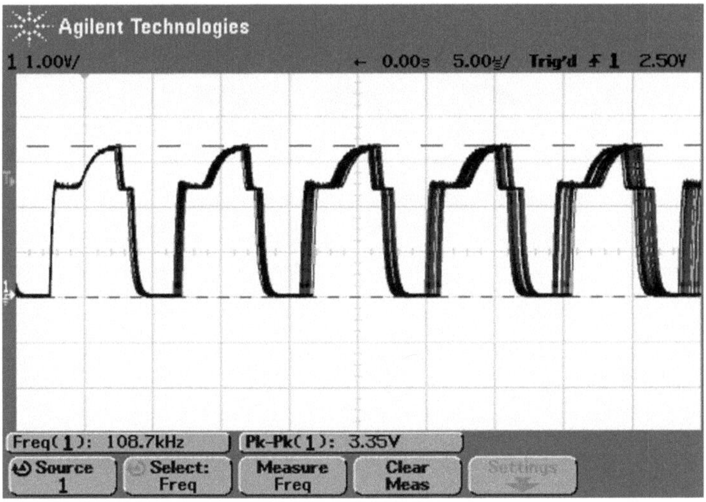

Fig. 6.20 PWM signal, output is 7.1V with input 4V

Figure 6.20 shows the PWM with the same microcontroller settings but the load has been reduced by 80%, thus the current drawn has increased from 158mA to 840mA and the voltage has dropped at the output from 8.1V to 7.1V. Thus the PWM is varying significantly more to try to correct this problem.

Figure 6.21 shows some of the internal variables of the program, the most important ones to watch to understand what is happening are *ADCAdj*, *GoalVoltage*, *CurrVoltage* and *Variance*.

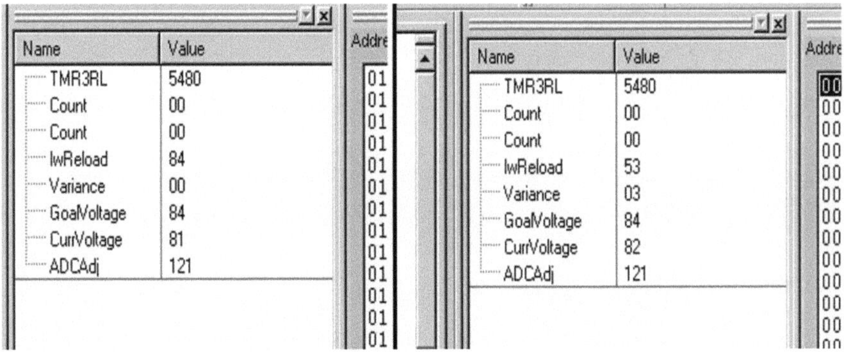

Fig. 6.21 Internal variables for the program

Figure 6.22 shows the PWM and the output voltage from the circuit. It can be seen that there is ripple on the output voltage.

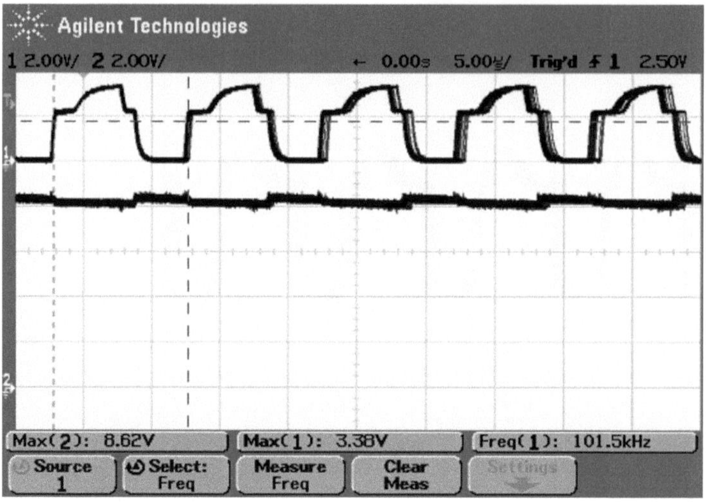

Fig. 6.22 PWM and O/P voltage

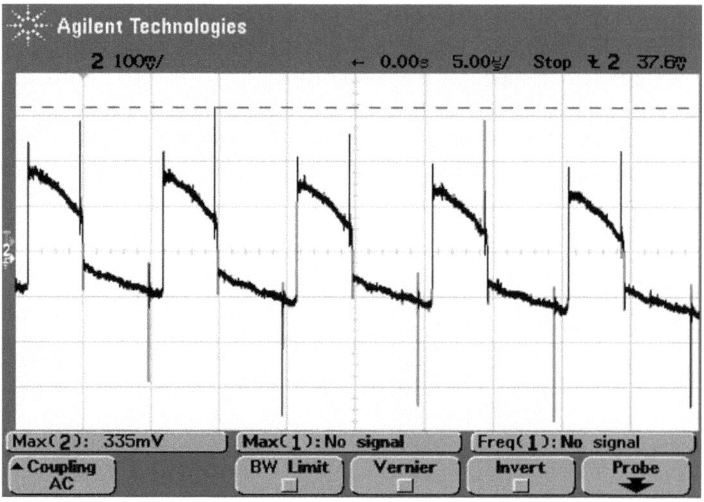

Fig. 6.23 Ripple on the output voltage

Figure 6.23 shows the output voltage ripple. The main ripple is approx 400mV p-p with spikes that go as high as 335mV and as low as -350mV. The oscilloscope was used in AC coupling mode to eliminate the DC offset of approx 8.4V.

7

Embedded Microcontroller Based Magnetic Levitation

7.1 Introduction

Magnetic levitation is a fundamental requirement for implementing a magnetic bearing (MB). In a magnetic bearing system, it provides a contact-free support for the rotating shaft. This is achieved through an attractive magnetic levitation force, produced by passing current(s) through electromagnet(s). The magnetic force is controlled with the help of adjustable current by implementing a control system. The gap between stator and rotor is measured with a position sensor and is used as a means of controlling the levitation force of the magnetic bearing. In this chapter the implementation of the control system for magnetic levitation based on an embedded microcontroller has been described. If the implementation of magnetic bearing system is done by using only electromagnets, it is known as Active Magnetic Bearing (AMB) system.

7.2 Background and Motivation

The use of magnetic bearing system in rotating machines has steadily increased, as an alternative replacement to conventional ball bearing. It provides a non-contact means of supporting the rotating shaft through an attractive magnetic levitation. Due to the non-contact nature of the bearing and rotor, it offers several advantages over mechanical bearing.

- Frictionless: Since magnetic bearing provides a contact-free support between the bearing and the rotor, frictional force is absent in the system.
- Extended machine life: Frictionless nature of the system extends the life of the machine as it is free from wear and tear.
- High speed operation: Because of the frictionless nature of operation there is no limit on the rotor speed of rotation.
- Lubrication free operation: Since there is no physical contact, lubrication of the system is not required and can be operated under environmental conditions that prohibit the use of lubricants.

G. Sen Gupta and S.C. Mukhopadhyay: Embedded Microcontroller Interfacing, LNEE 65, pp. 127–155.
springerlink.com

- Low losses and noiseless operation: mechanical frictional loss and frictional noise is reduced due to the non-contact nature of the system.
- Adjustable damping and stiffness characteristics due to use of electronics.
- Extremely reliable.

There are many applications that utilized MB's advantages over the traditional ball bearing. These applications can be found in different fields such as: industrial, military and space applications; vacuum, low and high temperature atmosphere; friction free operation for gyros, disc-drives; lubrication free operation for food-processing equipment, high vacuum pumps, clean-room machinery; and vibration isolation.

Inspite of all the advantages, AMB's are still too complicated and expensive. A few drawbacks are:

- The control circuits required are complicated
- A high precision displacement sensor is required to determine the rotor position
- Power Consumption: current required by the electromagnets result in high power consumption.

Generally speaking, a magnetic bearing system consists of magnetic actuators, controllers, amplifiers, and sensors. Unlike conventional bearings such as rolling element bearings and fluid film bearings, magnetic bearings require active control as they are unstable under open-loop operating condition. This instability also exists even when permanent magnets are used, since it is not possible to suspend an object in all axes at stable conditions. The rotor position is fed back to the controller via the position sensors. The controller sends a command signal to the amplifier which produces necessary currents in the actuator coils. Magnetic forces are generated by the currents in the coils. Employing an appropriate controller will stabilize the suspended rotor.

In magnetic bearing technology electromagnets produce the necessary magnetic flux. The magnetic flux Φ can be visualized by magnetic field lines. The magnetic field intensity H is linked to the flux density B, by

$$B = \mu_0 \mu_r H$$

Here, $\mu_0 = 4\,\pi\,*10^{-7}$ H/m is the absolute permeability of the vacuum, and μ_r is the relative permeability depending on the medium the magnetic field acts upon. μ_r equals 1 in a vacuum, as well as in air. By using ferromagnetic material, where μ_r is generally very large, the magnetic flux can be concentrated in the core material.

Figure 7.1 shows the forces that are found in an electromagnet. Φ is the magnetic flux, u is the coil voltage, i is the current passing through the coil, s is the gap between the suspended object and the electromagnet, and A is the cross-sectional area of the object.

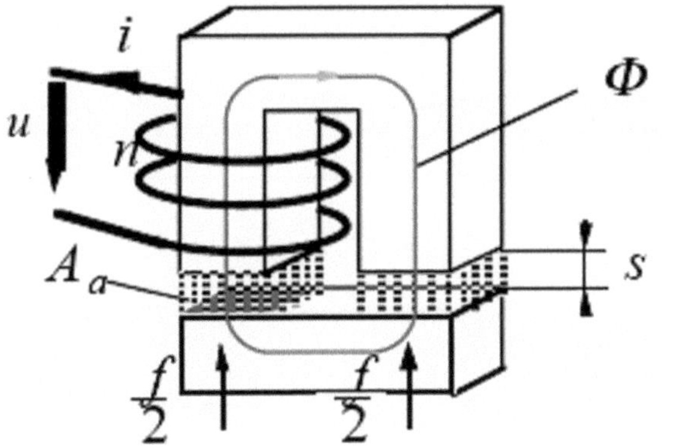

Fig. 7.1 Forces of an electromagnet

7.3 Hybrid Active Magnetic Bearing

The system that is used throughout this chapter is called a Hybrid Magnetic Bearing (HMB). It employs both passive and active ways of supporting the rotating shaft. It contains electromagnets, which are for active support, and permanent magnets found on the upper and lower part of the system, which provide passive support. The upper and lower permanent magnets found in HMB give not only support but help to reduce the load that the electromagnet has to carry, thereby minimizing the power loss. Figure 7.2 shows the different components of a hybrid magnetic bearing system.

A few components are described in the following sections:

7.3.1 Displacement Sensor

Sensors supply the information to the controllers. Either position or flux can be measured, but position sensing is more widely used. There are several different position sensors which includes the Hall sensor, capacitive sensor, and eddy current sensor. The eddy current sensors have the best characteristics in terms of bandwidth and phase shift.

The displacement sensor measures the gap between the shaft and the rotor in the vertical axis. The sensor outputs a voltage proportional to the gap which is fed to a control system. This control system compares the current gap value with a reference value and subsequently adjusts the current in the electromagnet accordingly to change the magnetic levitation of the shaft.

7.3.2 Permanent Magnet

The permanent magnets can be found in the upper and lower part of the system. They are used to provide stability and aid in supporting the weight of the shaft in the system. Figure 7.3 shows the fabricated magnetic bearing with 24 magnets each in the stator and the rotor. The disc is made of aluminium material. The 24

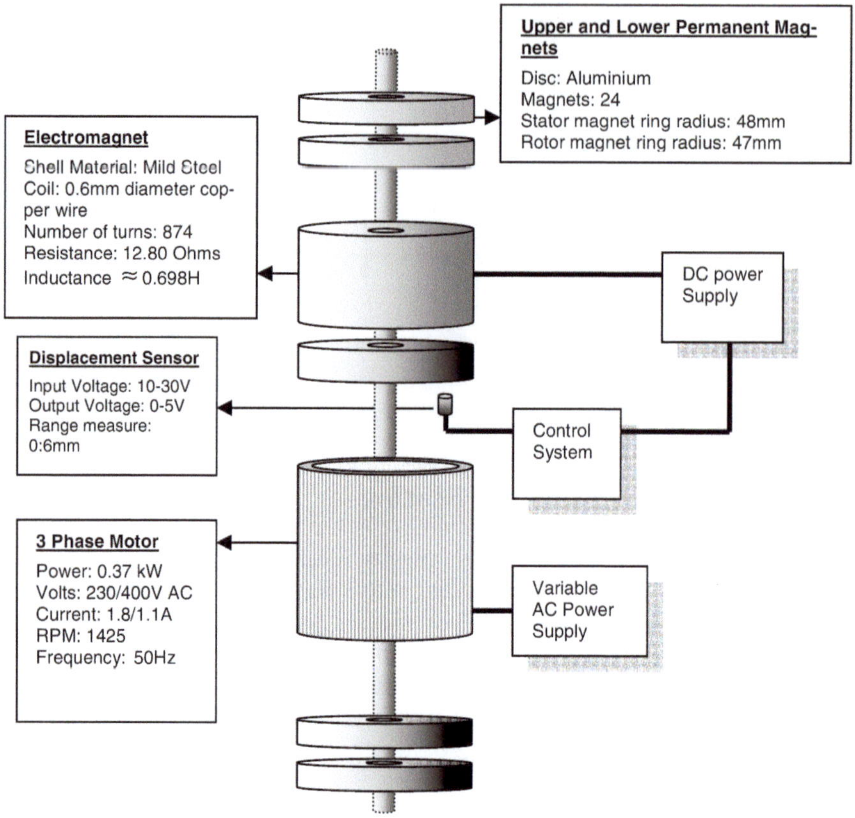

Fig. 7.2 Diagram showing the major components of the fabricated magnetic bearing system

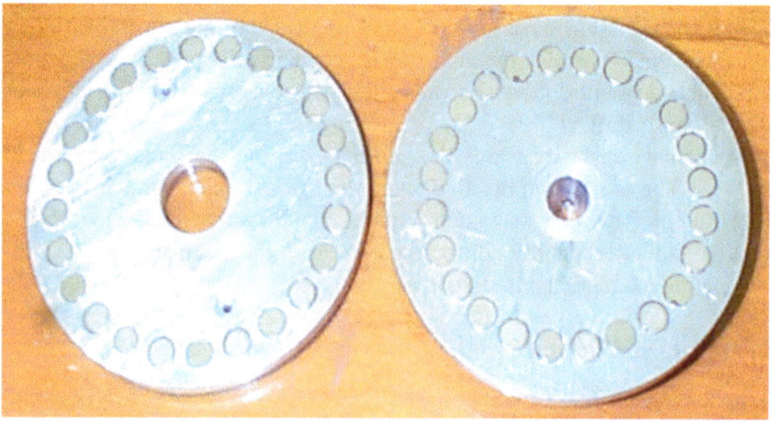

Fig. 7.3 Fabricated magnetic bearing with 24 magnets each in stator and rotor

magnets are placed around the aluminium disc; 48 mm radius for stator and 47 mm radius for rotor. This is a unique way of achieving magnetic levitation.

7.3.3 Electromagnet and Force Relationship

The electromagnet supports the weight of the system by providing a magnetic levitation force. The coil has 874 turns of copper wire and the diameter of the wire is 0.6mm. The electromagnet's resistance is 12.80 Ω and inductance is approximately 0.698 Henry. Table 7.1 shows the mass of the various components of the system.

Table 7.1 Mass calculations of the system

System components	Mass (kg)
Shaft and motor rotor	2.23
Permanent magnets	0.42
Electromagnet rotor	1.15
Total	3.80

It is assumed that the permanent magnets support 50% of the total weight. Since $F \equiv ma$, the total force that the electromagnetic requires in order to levitate the system is: 0.5*3.8*9.8 = 18.62 N. The electromagnet's force relationship with the gap and current is:

$$F = k\left(\frac{I_0}{x_0}\right)^2$$

At steady state of the system, I_0 and x_0 are the nominal current and gap values.

$$F = k\left(\frac{I_0 + \Delta i}{x_0 + \Delta x}\right)^2;$$ Δi and Δx are the deviation of current and gap

$$= k\frac{I_0^{\,2}}{x_0^{\,2}}\left(\frac{1+\dfrac{\Delta i}{I_0}}{1+\dfrac{\Delta x}{x_0}}\right)^2 = k\left(\frac{I_0}{x_0}\right)^2\left[\left(1+\frac{\Delta i}{I_0}\right)^2\left(1+\frac{\Delta x}{x_0}\right)^{-2}\right]$$

$$\approx k\left(\frac{I_0}{x_0}\right)^2\left[\left(1+2\frac{\Delta i}{I_0}\right)\left(1-2\frac{\Delta x}{x_0}\right)\right]$$

All higher order terms are neglected

$$= k\left(\frac{I_0}{x_0}\right)^2\left(1+2\frac{\Delta i}{I_0}-2\frac{\Delta x}{x_0}\right) = k\left(\frac{I_0}{x_0}\right)^2\left(1+\frac{2}{I_0}\Delta i-\frac{2}{x_0}\Delta x\right)$$
$$= F_0 + k_i\Delta i - k_s\Delta x$$

where,

$$F_0 = k\left(\frac{I_0}{x_0}\right)^2; \quad k_i = 2k\left(\frac{I_0}{x_0}\right)^2\left(\frac{1}{I_0}\right), \quad k_s = 2k\left(\frac{I_0}{x_0}\right)^2\left(\frac{1}{x_0}\right).$$

The nominal operating current is kept around 0.6 A and the maximum current is set at 1.0 A. The electromagnet and the displacement sensor are shown in figure 7.4. The electrical equation of the electromagnet around nominal gap:

$$\mathrm{u} = \mathrm{Ri} + \mathrm{L}\frac{\mathrm{di}}{\mathrm{dt}}\text{; u is the applied voltage across the coil}$$

Fig. 7.4 Electromagnet and displacement sensor

Figure 7.5 shows the picture of the fabricated hybrid magnetic bearing system.

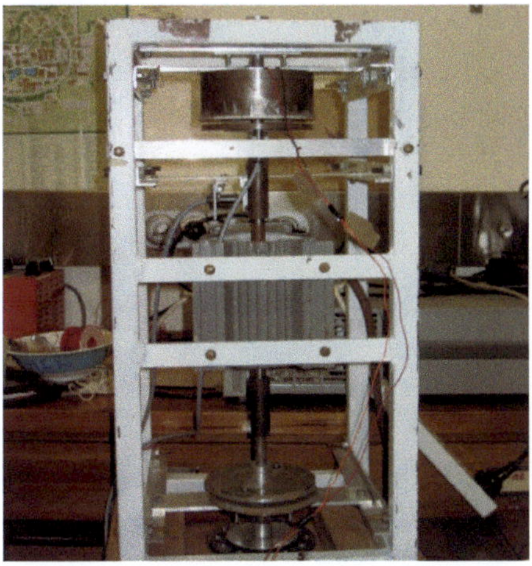

Fig. 7.5 Fabricated hybrid magnetic bearing system

7.4 Design of Control System

There are two types of control systems implemented in the AMB system. These are: (i) Analogue control system and (ii) embedded microcontroller-based control-system. Both systems employ a PID controller for optimum system performance.

7.4.1 PID Controller

Transfer function for a PID controller is:

$$K_p + \frac{K_i}{s} + K_d s = \frac{K_d s^2 + K_p s + K_i}{s}$$

where,			
	K_p	-	Proportional gain
	K_i	-	Integral gain
	K_d	-	Derivative gain

Proportional, Integral, and Derivative (PID) controllers appear to be the most widely used. The reason for the popularity can be attributed to the simplicity of implementation and robustness. In a closed-loop (CL) system, the characteristic of proportional (P), integral (I) and differential (D) controllers are given in table 7.2.

Table 7.2 Characteristics of the P, I and D controllers in a closed-loop system

CL response	Rise time	Overshoot	Settling time	Steady state error
K_p	decrease	increase	small change	decrease
K_i	decrease	increase	increase	eliminate
K_d	small change	decrease	decrease	small change

A proportional controller (K_p) has the effect of reducing the rise time but it can-never eliminate the steady-state error. However, employing an integral control (K_i) will eliminate the steady-state error but it affects the transient response. A derivative control (K_d) will have the effect of increasing the stability of the system, reducing the overshoot, and improving the transient response of the system. The values of the gains are adjusted to provide the desired performance of the system.

7.4.2 Analog Control System

To appreciate the advantages and usefulness of digital control, analog control has been implemented first. The implementation of an analog control system uses a LM324N operational amplifier (Op-Amp). In the chapter appendix the final schematic of the analog control system has been provided. Looking at each individual op-amp of the schematic, the first op-amp (U1A) shown in figure 7.6 is a unity gain inverter. It takes in the voltage signal from the displacement sensor and inverts it, since it is a negative feedback closed-loop control system.

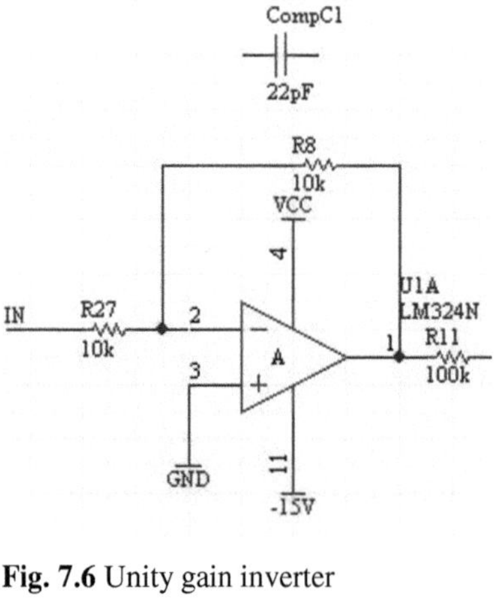

Fig. 7.6 Unity gain inverter

The second Op-Amp (U1B), shown in figure 7.7, is a non-inverting summing amplifier. It takes two signals: (i) the inverted displacement sensor voltage signal and

(ii) a reference signal, which corresponds to a desired constant gap for optimum system operation. The two signals are added together and the sum is the error signal, which is relayed to the PI and PD controllers.

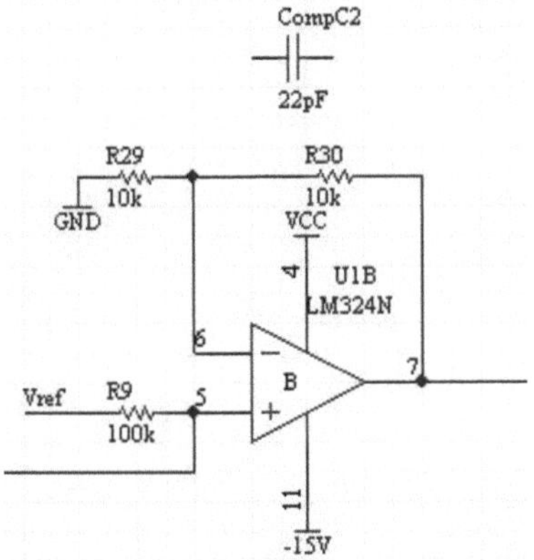

Fig. 7.7 Non-inverting amplifier with Vref and inverted gap sensor signal as input

The third Op-Amp (U1C), shown in figure 7.8, is a PI (proportional and integral) controller. It is a combination of an integrator circuit and a simple proportional gain. The combination of the feedback capacitor C_1 and resistor R_{36} to the non-inverting input presents the integral control. The resistor R_3 and the feedback resistor R_{36} provide the proportional control. The transfer function of an inverting Op-Amp with resistor R3 as the input element and resistor R_{36} and capacitor C_1 as a feedback element is:

$$G(s) = \frac{-R_{36}}{R_3}\left(\frac{s + 1/R_{36}C_1}{s}\right)$$

Pole at the origin and a zero at $-1/R_{36}C_1$

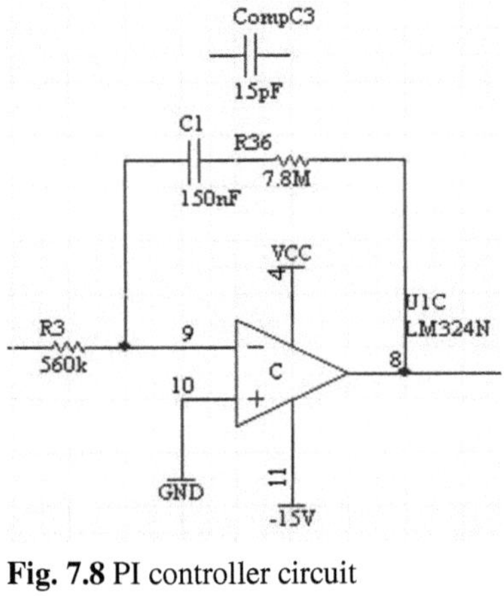

Fig. 7.8 PI controller circuit

The fourth Op-Amp U1D, shown in figure 7.9, is a PD (proportional and derivative) controller. The combination of the capacitor C_2 and the feedback resistor R_7 at the non-inverting input presents derivative control. The resistor R_{12} and the feedback resistor R_7 provide the proportional control.

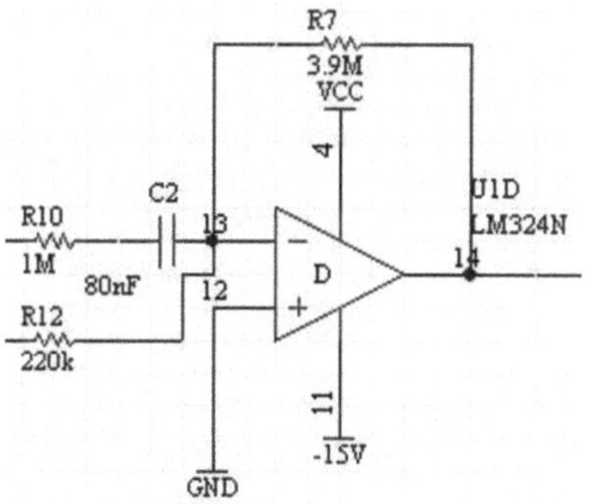

Fig. 7.9 PD controller circuit

Figure 7.10 shows the combination of the two controllers, PD and PI. The input for both the controllers is the error signal from figure 7.7. The combined output of the PD and PI controller provides a desired response such as eliminating the steady-state error, reducing the overshoot, improving the transient response and increasing the stability of the system.

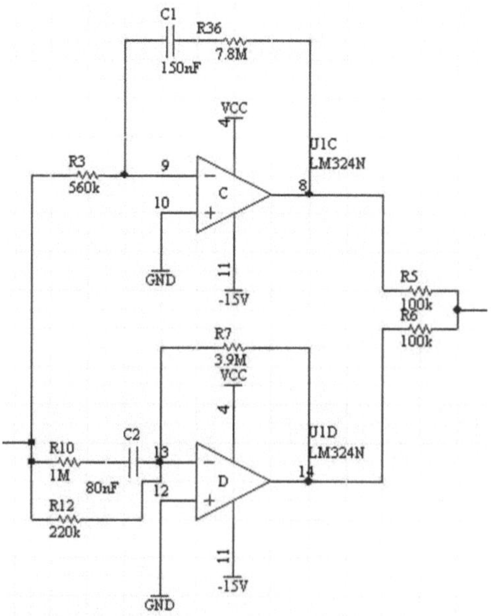

Fig. 7.10 PI and PD controller

The fifth Op-Amp U2, shown in figure 7.11 is a non-inverting amplifier with a negative feedback. The Op-Amp takes both output signal of PI and PD controller as a non-inverting input. It combines the two signals and gives an overall effect of a PID controller.

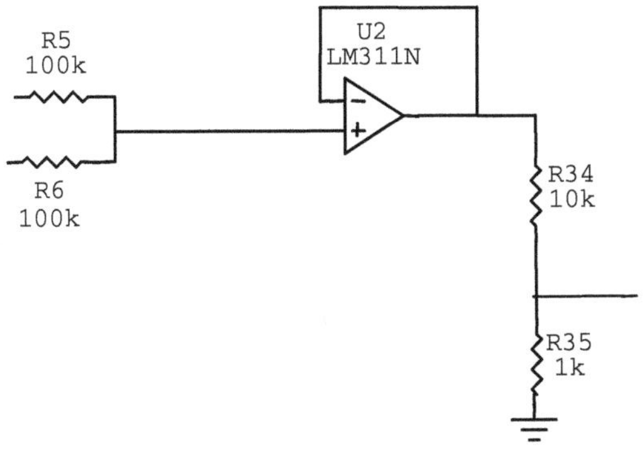

Fig. 7.11 Non-inverting amplifier with scaling down voltage divider network

A part of the output of the final Op-Amp is fed into an external DC power supply, which applies the proportional current into the electromagnet coil for system stability at constant optimum gap.

7.4.3 Results from the Controller

An oscilloscope is used to see the waveform of the rotor position as an outcome of the analog control. There are two signals which have been looked at: (i) Actual signal and (ii) the control signal to the power supply. In all the waveforms, the top signal corresponds to the actual gap and the bottom signal corresponds to the control signal to the power supply. Figure 7.12 shows the steady state response of the system during normal operation.

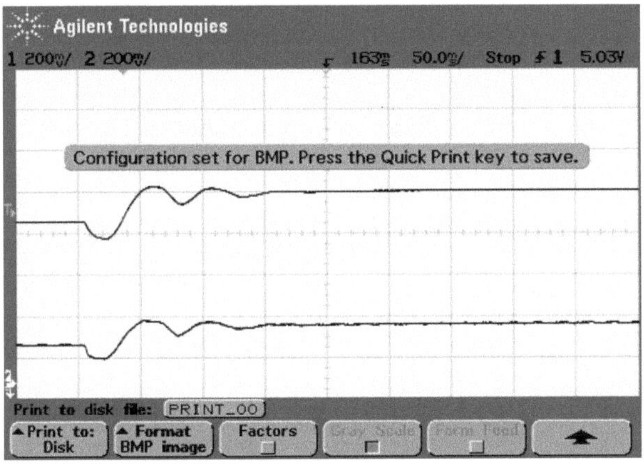

Fig. 7.12 Steady state response during normal operation

In addition, some disturbances were introduced to see how it affects the performance of the system. Figure 7.13 shows the waveform changes of the system during disturbances.

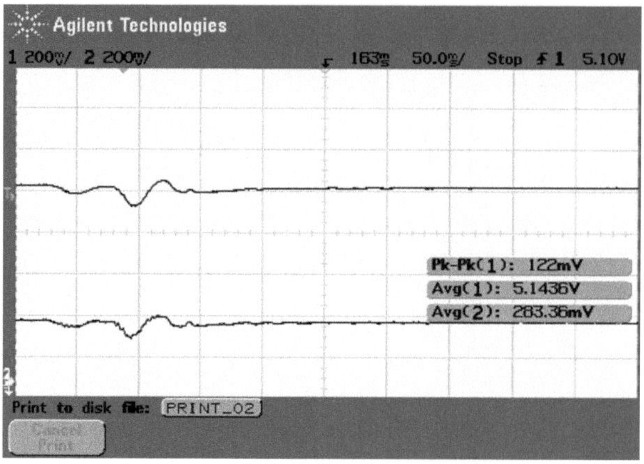

Fig. 7.13 Vibration during disturbance

Figure 7.14 shows the steady state and transient response of the system, when K_d is increased. The image from the left shows the steady state response of the system from starting and the right image is the transient (disturbance) response, when K_d is increased.

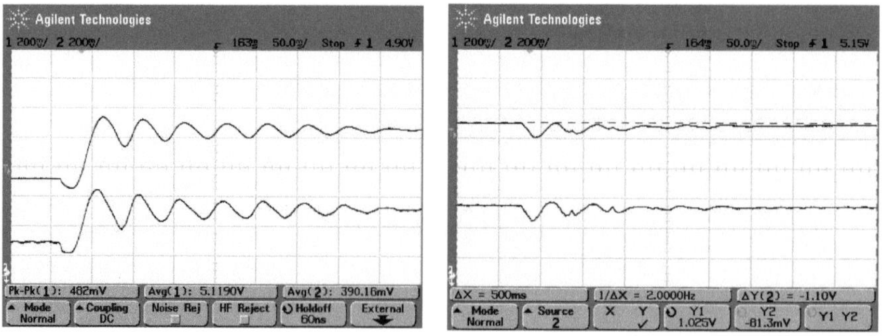

Fig. 7.14 Steady-state and transient response with increased *Kd* value

Figure 7.15 shows the steady state and transient response of the system when K_d is decreased. If the results of figures 7.14 and 7.15 are compared it is seen that an increase in K_d value decreases the overshoot and settling time of the system.

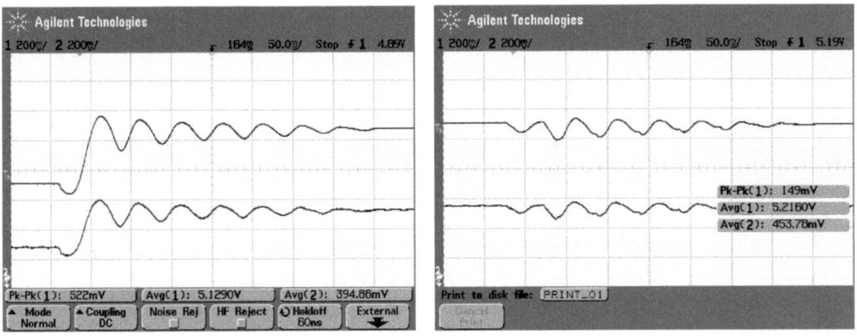

Fig. 7.15 Steady-state and transient response with decreased *Kd* value

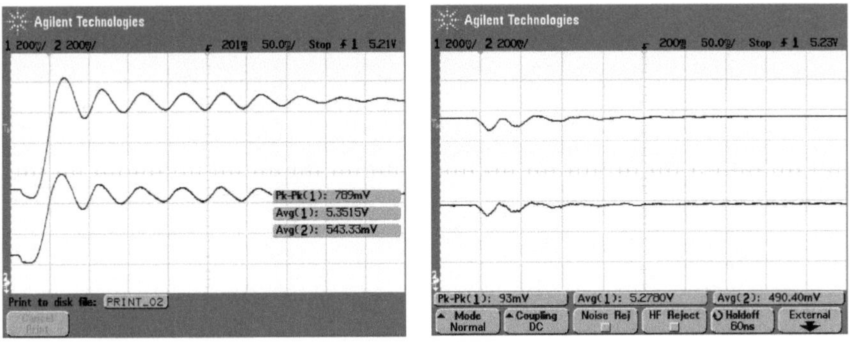

Fig. 7.16 Steady-state and transient response with decreased *Ki* value

Figure 7.16 shows the steady state and transient response of the system when K_i is decreased. The left image is the steady state response from starting and right image is the transient (disturbance) response.

Furthermore, figure 7.17 displays the steady state and transient response of the system when K_i gain value is increased. The left image in figure 7.17 is the steady state response and on the right is the transient (disturbance) response. Comparing the two figures (figures 7.16 and 7.17), when the K_i gain value is increased there is an increase in overshoot.

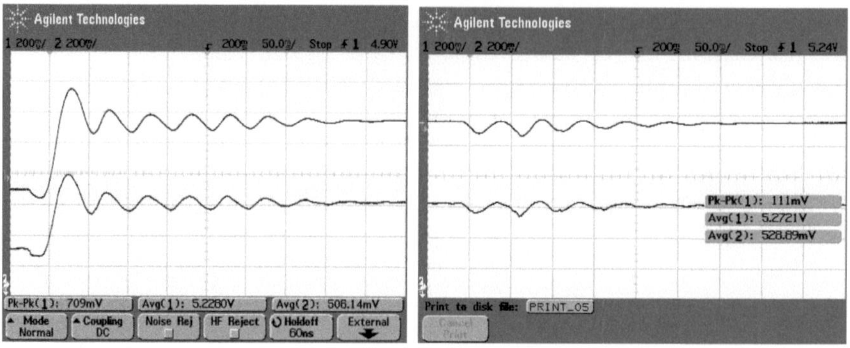

Fig. 7.17 Steady-state and transient response with increased *Ki* value

The steady state and transient response of the system is shown in figure 7.18 when the proportional gain K_p is decreased. The left image of figure 7.18 is the steady state response and on the right image is the transient (disturbance) response when K_p is decreased.

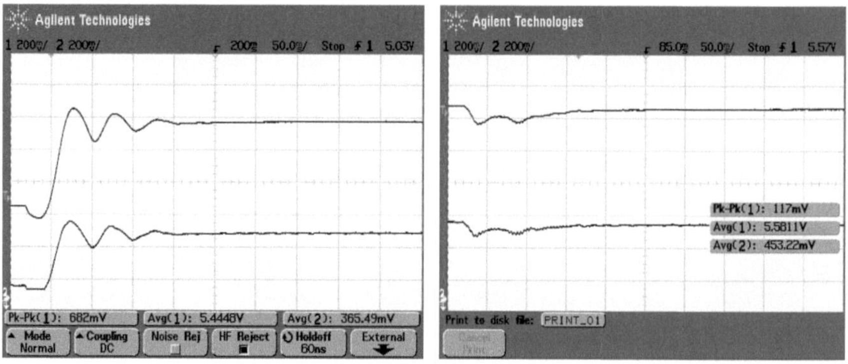

Fig. 7.18 Steady-state and transient response with decreased *Kp* value

Figure 7.19 shows the system's steady state and transient response when K_p is increased. The left part of figure 7.19 is the steady state response and on the right is the transient (disturbance) response. If the figures 7.18 and 7.19 are compared, it is seen that the rise time and steady state error is decreased when K_p value is increased.

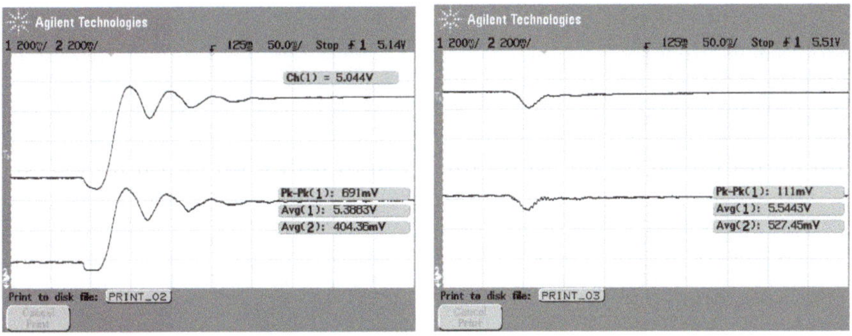

Fig. 7.19 Steady-state and transient response with increased *Kp* value

7.5 Microcontroller Based Control System

With the above knowledge and understanding of the analog control system, it will be much easier to understand the digital control. The microcontroller board, as shown in figure 7.20, has been used to provide the control system. The expansion board contains the potentiometer (for reference adjustment), LCD (for displaying the actual gap and reference gap) and the sensor input, where the signal from the displacement sensor is connected. As far as the external interfacing is concerned the analog displacement signal goes to the analog input, shown as sensor input in figure 7.20, and the PWM output is taken from port 1 (Port 1.0), shown in figure 7.20 as PWM output.

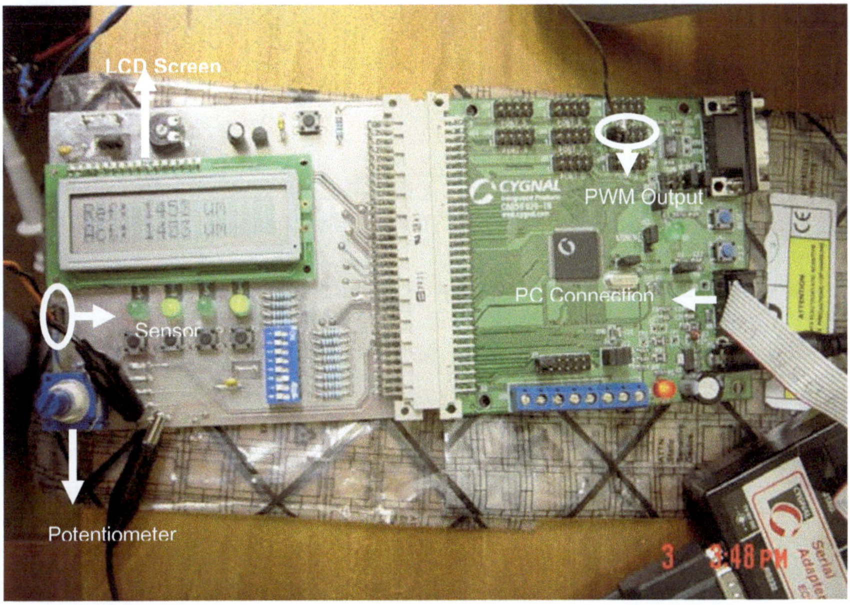

Fig. 7.20 Microcontroller Development Board with Expansion Board

Pulse Width Modulation (PWM) signal is generated in the microcontroller board, and it produces a square waveform with a variable on-to-off ratio. The PWM signal controls the on/off state of the transistor, either allowing the current to flow through the electromagnet or limiting it. The magnetic bearing system is an unstable system and the use of PID controller is implemented to provide stability of the system. The overall system setup for the microcontroller based control is shown in figure 7.21.

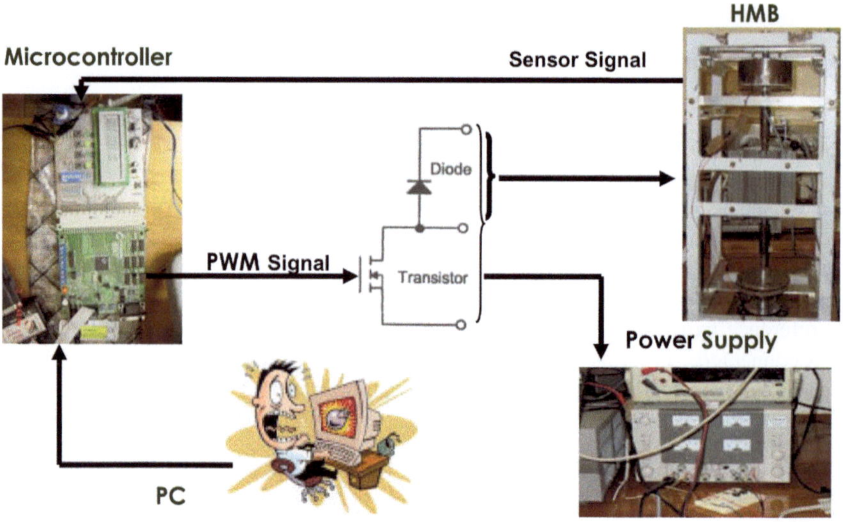

Fig. 7.21 Equipment Setup for microcontroller-based control

The displacement signal is taken as an input variable and goes through the expansion board of the microcontroller. The personal computer (PC) modifies the program to implement different gain values of K_p, K_i and K_d. It uses a RS232 serial cable to connect from the serial port (COM) of the PC to the microcontroller which is also used to download the complied code. The reference signal, which corresponds to the desired constant gap for optimum system operation, is adjusted using the potentiometer in the expansion board. Both the signals, reference and sensor, are displayed on the LCD. The unit for the two displayed signals is micrometer. The PWM output, located at Port1.0 of the development board (also known as target board), is fed through the transistor switch. Figure 7.22 shows the topography of the target board. The PWM signal turns on and off the switch, hence controlling the electromagnet's current. A 3-output power supply is used to provide the voltage to run the control system. A PID controller has been implemented for good performance and system stability.

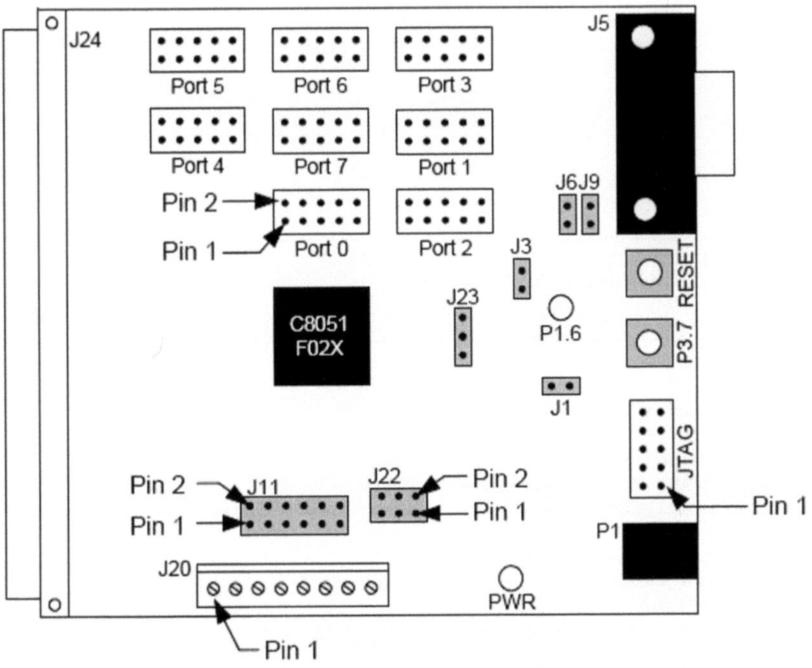

Fig. 7.22 C8051F020 Target Board

7.5.1 Microcontroller Code

Timer 0 was used to generate the PWM signal. *MAX_Count* is set to 255. Each time the Timer 0 interrupt service routine is executed, *PWM_counter* is incremented by 1. When the *PWM_counter* value exceeds the threshold value (0 to 255) set by the potentiometer (*dutyCycleCount*), the PWM output will be reset. When the *PWM_counter* reaches the *MAX_Count*, the output is set to 1 and counter is reset. 22.1148 MHz external oscillator was used. Figure 7.23 shows the ISR (Interrupt Service Routine) of Timer 0 that sets the PWM output.

```
void Timer0_ISR (void) interrupt 1
{
        //-- clear TF0
        TF0 = 0;
        PWM_counter++;
        if (PWM_counter >= dutyCycleCount)
        {
                PWM_output = 0;
                if (PWM_counter >= MAX_Count)
                {
                        PWM_output = 1;
                        PWM_counter = 0;
                }
        }

}
```

Fig. 7.23 Timer0_ISR code for PWM_counter

```
while(1)
{
  current_error_gap = Ref - Act;

  //For PID control, Gain Values are:
  // Kp=55/10; Ki = 6/10; Kd = 22/100;
  proportional_factor = (current_error_gap * 55)/10;
  integral_factor = integral_factor +
        (current_error_gap*60)/100;
  derivative_factor = ((current_error_gap -
        previous_error_gap)*22)/100;
   previous_error_gap = current_error_gap;

  //Integral factor limits at 10 <= integral_factor<= 100
  if (integral_factor > 100) integral_factor=100;
  if (integral_factor < 10) integral_factor=10;

  //actual and reference gap in micrometers
  gap_ref = ((33000*Ref)/(4095*8));
  gap = ((33000*Act)/(4095*8));

  //Displays the reference and actual values into the lcd
  lcd_clear();
  printf("Ref: %4d um", gap);
  lcd_goto(0x40);
  printf("Act: %4d um", gap_ref);

  dutyCycleCount = proportional_factor + integral_factor +
        derivative_factor;
  // DutyCycle limited at 40/255 (15% <= dutyCycleCount <= 95%)
  if (dutyCycleCount < 40) dutyCycleCount = 40;
  if (dutyCycleCount > 240) dutyCycleCount = 240;
  large_delay(1);
}
```

Fig. 7.24 Program for factor and duty cycle calculations in microcontroller control

The code for implementing the PID controller for the system is shown in figure 7.24. It also shows the PID gain values and the *dutyCycleCount* calculations. The *current_error_gap* calculation is the difference between the reference signal, set by the potentiometer, and the actual signal, coming from the displacement sensor.

The PID factors are calculated with respect to the PID gain values. The proportional factor is the product of the current error and proportional gain (K_p) while the integral factor is the sum of the previous integral factor plus the product of the error and integral gain (K_i). For derivative factor, it is the difference between the previous gap error and the current gap error multiplied by the derivative gain (K_d). The *dutyCycleCount* is the sum of the three calculated factors. However, the *dutyCycleCount* is restricted to 10% to 95% operation. Also, the integral factor was limited to operate between 10 and 100. The *dutyCycleCount*, together with the *PWM_counter*, sets the PWM output. The compiler used doesn't support floating point operations. So it is important to have all multiplication done first before any division takes place.

ADC0 was used to convert the two signals - sensor signal input and the reference signal from the potentiometer. The displacement sensor signal is connected to the input pin AIN.3 and the reference signal is connected to AIN.2. Initially, the *ChannelFlag* was set to 1 and then the signals are processed alternately. Figure 7.25 shows the interrupt service routine of the ADC. Figure 7.26 shows the program structure of the overall control system.

```
void ADC0_ISR(void) interrupt 15 using 1
{

        AD0INT = 0;     // clear ADC Conversion

        //Selects one of the two signal - reference and actual
        switch (ChannelFlag)
        {
                case 0:
                        Act = ADC0;
                        ChannelFlag = 1;
                        AMX0SL = 0x03; // Select Analog input
                                       // AIN.3- actual gap
                        AMX0CF = 0x00; //     "
                        break;
                case 1:
                        Ref = ADC0;
                        ChannelFlag = 0;
                        AMX0SL = 0x02; // Select Analog input
                                       // AIN.2- POT reference gap
                        AMX0CF = 0x00; //     "
                        break;

        }
}
```

Fig. 7.25 ADC0_ISR code for the two signals – reference and actual

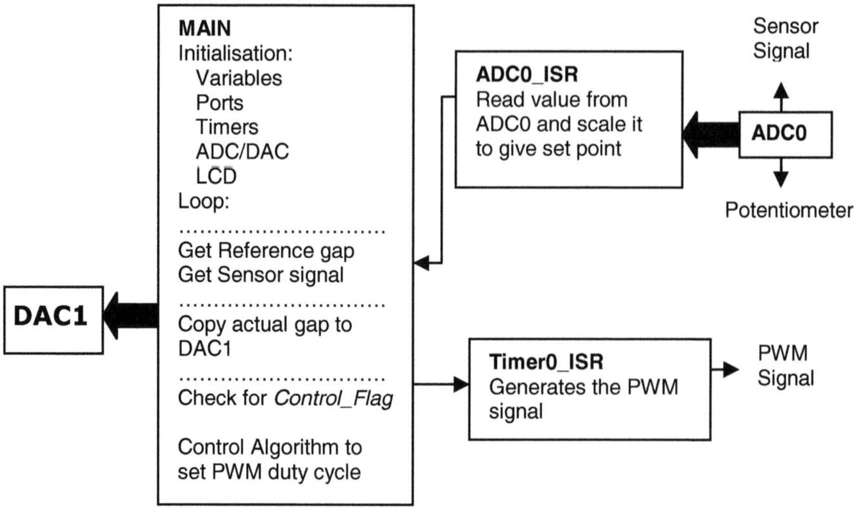

Fig. 7.26 Program structure

7.5.2 Results of the Microcontroller Based Control

The output waveform of the microcontroller based system was observed on an oscilloscope. Figure 7.27 shows the output of a PD controller. The top waveform shows the steady state response for the system during normal operation. The reference and gain values are:

- $K_p = 3.0$
- $K_d = 0.1$
- Ref gap = 1.523 mm

The waveform in figure 7.28 shows the output of the system with PID control. Comparing figures 7.27 and 7.28, the output waveform with PID control is a lot better with no overshoots or oscillations. Introducing the integral control improves the overshoot, steady state and transient response of the system. The PID gain and reference values used are shown below.

- $K_p = 3.0$
- $K_i = 0.1$
- $K_d = 0.5$
- Gap Ref = 1.523 mm

The reference signal is set to 1.523 mm from the potentiometer. K_p, and K_d values are set the same as the one in PD control and K_i value is set at 0.1.

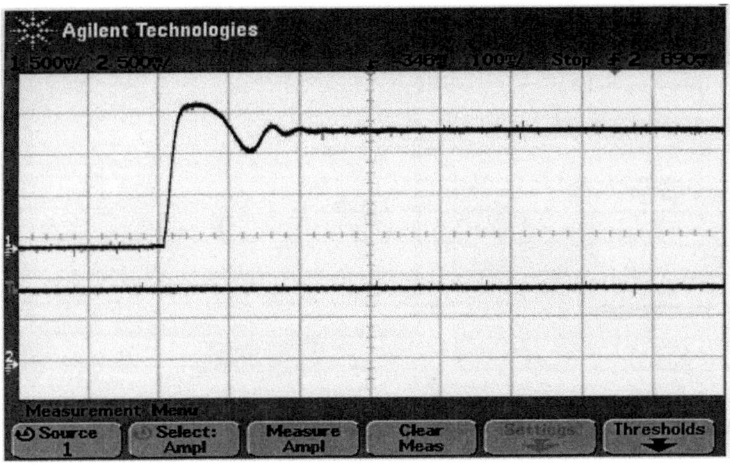

Fig. 7.27 Microcontroller output signal with PD control

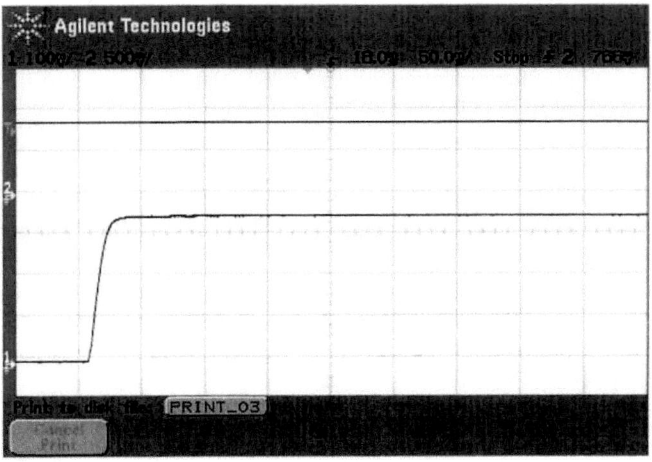

Fig. 7.28 Microcontroller output signal with PID control

Figure 7.29 shows the steady-state and transient response of the control system during normal operation. Figure on the left is the steady-state operation of the system and on the right is the effect of the disturbances in normal operation. Disturbance is introduced by tapping the HMB while it is running.

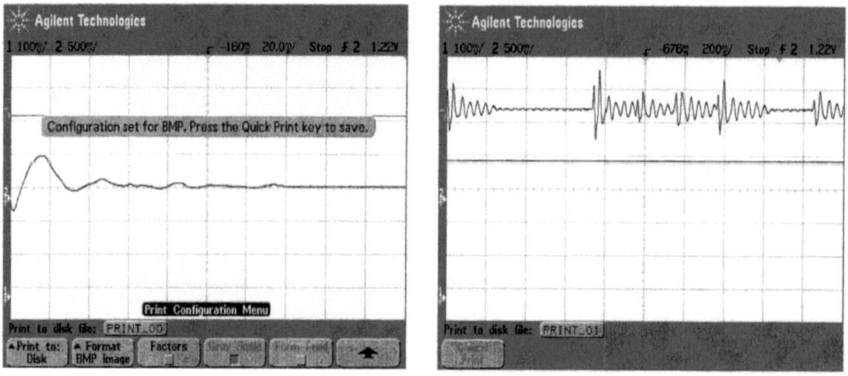

Fig. 7.29 Steady-state and transient response during normal operation

7.6 Conclusions

This chapter has described two types of control systems implemented for the Hybrid Magnetic Levitation system. The two controllers employ a PID control to provide system stability. The first controller is an analog control system. For this controller, it uses a LM324N, operational amplifiers. Each op-amp does a specific task to provide the overall system control. The major drawback of this control is the inflexibility of changing the gain values. Since the control is hardware-based, changing the gain values involves changing the components (i.e. resistor and capacitor value). Having to do this whenever you want to change the gain value is time consuming. The second controller is a microcontroller-based control. Unlike the analog system, the microcontroller provides easy ways of changing the gain values. It is a software based control; therefore changing the gain values requires you to just change the value of the gain variable in the program. Furthermore, the microcontroller based control system is more accurate than the analogue; it is possible to set any desired P, I and D gain values for optimum performance.

A7 Chapter Appendix

A7.1 Microcontroller Code Listing

```
// This program is used to test the various parts of the expansion
// board.It takes input from the serial port to select each test.

//-----------------------------------------------------------------
// Includes
//-----------------------------------------------------------------
#include <stdio.h>
#include <c8051f020.h>                    // SFR declarations
//-----------------------------------------------------------------
// 16-bit SFR Definitions for 'F02x
//-----------------------------------------------------------------
sfr16 DP              = 0x82;             // data pointer
sfr16 ADC0            = 0xbe;             // ADC0 data
sfr16 ADC0GT          = 0xc4;             // ADC0 greater than window
sfr16 ADC0LT          = 0xc6;             // ADC0 less than window
sfr16 DAC0            = 0xd2;             // DAC0 data
sfr16 DAC1            = 0xd5;             // DAC1 data

//-----------------------------------------------------------------
// Global DEFINES
//-----------------------------------------------------------------
#define uchar unsigned char

#define SYSCLK         22118400         // system clock frequency in Hz
#define LCD_DAT_PORT   P6               // LCD is in 8 bit mode
#define LCD_CTRL_PORT  P7               // 3 control pins on P7
#define RS_MASK        0x01             // for assessing LCD_CTRL_PORT
#define RW_MASK        0x02
#define E_MASK         0x04

#define DIP    P4                       // DIP switches
#define PB     P5                       // Push buttons bit 0 - 3
#define LED    P5                       // LEDs bit 4 -7

#define MAX_Count 256

//-----------------------------------------------------------------
// Global MACROS
//-----------------------------------------------------------------
#define pulse_E();\
       small_delay(1);\
       LCD_CTRL_PORT = LCD_CTRL_PORT | E_MASK;\
       small_delay(1);\
       LCD_CTRL_PORT = LCD_CTRL_PORT & ~E_MASK;\

//-----------------------------------------------------------------
// Global CONSTANTS
//-----------------------------------------------------------------
sbit SW2 = P3 ^ 7;
sbit PWM_output = P1^0;                 //-- PWM output pin

unsigned int SAMPLE_RATE = 2;           // sampling rate - 2 Hz
unsigned int Ref;                       //Reference
unsigned int Act;                       //Actual signal
unsigned int ChannelFlag;
```

```
int current_error_gap;                  //Current error gap
int proportional_factor;                //Proportional Factor
int integral_factor;                    //Integral Factor
int derivative_factor;                  //Derivative factor

char control_action;

unsigned int gap;
unsigned int gap_ref;
int previous_error_gap;
unsigned int PWM_counter;
int dutyCycleCount;

//--------------------------------------------------------------------
// Declaration of functions.
//--------------------------------------------------------------------
void init(void);
void Init_Port(void);                   //-- Configures the Crossbar and
GPIO ports
void lcd_init(void);
void Init_Clock(void);
void Init_ADC0(void);
void Init_Timer0 (unsigned char reload);
void Timer0_ISR (void);
void Init_DAC1(void);

void lcd_goto(uchar addr);              // move to address addr
void lcd_busy_wait  (void);             // wait until the lcd is no
longer busy
char putchar(char dat);
void lcd_cmd(char cmd);
void lcd_clear(void);
void lcd_cursor(bit on);

void ADC0_ISR(void);
void error_msg(int x);
void small_delay    (char d); //8 bit,about 0.34us per count @22.1MHz
void large_delay    (char d); // 16 bit,about 82us per count @22.1MHz
void huge_delay     (char d); // 24 bit,about 22ms per count @22.1MHz

void main(void)
{
        EA = 0;                 // Disable global interrupts
        PWM_counter = 0;
        current_error_gap = 0;
        proportional_factor = 0;
        integral_factor = 0;
        derivative_factor = 0;

        previous_error_gap = 0;
        gap = 0;
        gap_ref = 0;

        lcd_init();
        init();
        Init_Clock();  // initialize the system clock
        Init_Port();
        Init_Timer0(79);
        lcd_cursor(0);
```

```
        init_timer3(SYSCLK/SAMPLE_RATE); // initialize Timer3 to
                                         // overflow at sample rate
        EA = 1;
        ChannelFlag = 1;
        Init_ADC0();
        Init_DAC1();

        while(1) {
                current_error_gap = Ref - Act;

                //For PID control, Gain Values are:
                // Kp=55/10; Ki = 6/10; Kd = 22/100;
                proportional_factor  = (current_error_gap * 100)/10;
                integral_factor = integral_factor +
                                (current_error_gap*10)/100;
                derivative_factor = ((current_error_gap -
                                previous_error_gap)*100)/1000;
                previous_error_gap = current_error_gap;

                //Proportional gain limits at 10 <= Kp <= 100
                if (integral_factor > 100) integral_factor=100;
                if (integral_factor < 10) integral_factor=10;

                //actual and reference gap in micrometres
                gap_ref = ((33000*Ref)/(4095*8));
                gap = ((33000*Act)/(4095*8));

                //Displays the ref and actual values into the LCD
                lcd_clear();
                printf("Ref: %4d um", gap);
                lcd_goto(0x40);
                printf("Act: %4d um", gap_ref);

                dutyCycleCount = proportional_factor +
                                integral_factor + derivative_factor;
                // DutyCycle limited at 40/255 (15% <= dutyCycleCount
                // <=  95%)240/255
                if (dutyCycleCount < 40) dutyCycleCount = 40;
                if (dutyCycleCount > 240) dutyCycleCount = 240;
                large_delay(1);
        }
}

//---------------------------------------------------------------------
// init - general initialization
//---------------------------------------------------------------------
void init(void)
{
        WDTCN = 0x07;           // Watchdog Timer Control Register
        WDTCN = 0xDE;           // Disable watch dog timer
        WDTCN = 0xAD;
}

void Init_Clock(void)
{

        OSCXCN = 0x67; // Crsytal Osc. Mode without divide by 2 stage
                // External Osc Freq Control Bits (XFCN2-0) set to 111
                // because crystal frequency > 6.7 MHz
```

```
        // wait till XTLVLD pin is set i.e., crystal is stabilized
        while ( !(OSCXCN & 0x80) );

        OSCICN = 0x88;          // 1000 1000b
        // Bit 2:Internal Osc. disabled (IOSCEN = 0)
        // Bit 3:Uses External Oscillator as System Clock (CLKSL = 1)
        // Bit 7:Missing Clock Detector Enabled (MSCLKE = 1)
}

void Init_Port(void) // Configures the Crossbar and GPIO ports
{
        P74OUT |= 0x08; // Setting the P5H and P5L to
                        // push-pull operation
        LED &= 0xF0;    // LEDs are turned OFF
        P2MDOUT = 0xFF; // Enable Port2 as (Push-Pull) output
        XBR0 = 0x04;    // Enable UART0
        XBR1 = 0x04;    // Enable INT0 (Pin P0.2)
        XBR2 = 0x40;    //Enable Crossbar and weak pull-ups (globally)
        P0MDOUT |= 0x00; // Enable Port 0 as a open drain output

        P1MDOUT = 0x01; //-- P1.0 Push-Pull
        P2MDOUT = 0x00;// Output configuration for P2
        P3MDOUT = 0x00;

}

void Init_Timer0 (unsigned char reload)
{
        CKCON = 0xF8;   //-- Use system clock (T0M = 1)
        TMOD = 0x02;    //-- Timer 0 in Mode 2 and incremented by
                        //   clock defined by T0M
        TL0 = 0xFF;     //-- Set to reload immediately
        TH0 = reload;   //-- Reload value (can be 0 to 255)
        ET0 =1;         //-- Enable Timer 0 interrupts
        TR0 = 1;        //-- Start Timer 0
}

//Used to generate the PWM
void Timer0_ISR (void) interrupt 1
{
        //-- clear TF0
        TF0 = 0;

        PWM_counter++;
        if (PWM_counter >= dutyCycleCount)
        {
                PWM_output = 0;
                if (PWM_counter >= MAX_Count)
                {
                        PWM_output = 1;
                        PWM_counter = 0;
                }
        }
}

void Init_ADC0(void)
{
        REF0CN |= 0x03;         // Enable internal bias generator and
                // internal reference buffer which gives 2.4 V ref
```

```
                // Select ADC0 reference from VREF0 pin
        ADC0CF = 0xB0;  // SAR0 conversion clock = 961 kHz was 0x08
                        // with Gain = 1

        //AMX0SL = 0x03;  // Select Analog input AIN.3- actual gap
        //AMX0CF = 0x00;  //     "
        ADC0CN = 0x84; // enable ADC0, Continuous Tracking
                        // Mode Conversion initiated on Timer 3
                        // overflow, ADC0 data is right justified
        EIE2 |= 0x02;           // enable ADC interrupts

}

void ADC0_ISR(void) interrupt 15 using 1
{
        AD0INT = 0;                     // clear ADC Conversion
                                        // complete indicator

        //Selects one of the two signal - reference and actual
        switch (ChannelFlag) {
        case 0:
                Act = ADC0;
                ChannelFlag = 1;
                AMX0SL = 0x03; //Select input AIN.3- actual gap
                AMX0CF = 0x00; //     "
                break;
        case 1:
                Ref = ADC0;
                ChannelFlag = 0;
                AMX0SL = 0x02; //Select input AIN.2- POT reference gap
                AMX0CF = 0x00; //     "
                break;
        }
}

// Digital to Analogue
void Init_DAC1(void)
{
        REF0CN |= 0x12;         // Enable internal bias generator and
                                // internal reference buffer
        DAC1CN = 0x88; // enable DAC1, right justified and
//      DAC1 = Ref;    // DAC output updates on timer 3 overflow
}

void error_msg(int x) // Error directory for various tests
{
        switch (x) {
        case 0:
                lcd_clear();
                printf("ERROR 20: DIP");
                lcd_goto(0x40);
                printf("test failed");
                huge_delay(50);
                break;
        case 1:
                lcd_clear();
                printf("ERROR 21: Push");
                lcd_goto(0x40);
                printf("button failed");
```

```
                huge_delay(50);
                break;
        }
}
//-------------------------------------------------------------------
// lcd_goto
//-------------------------------------------------------------------
// change the text entry point
//
void lcd_goto(char addr)
{
        lcd_cmd(addr | 0x80);
}

#pragma OPTIMIZE (6)
void lcd_init(void)
{
        LCD_CTRL_PORT = LCD_CTRL_PORT & ~RS_MASK;    // RS = 0
        LCD_CTRL_PORT = LCD_CTRL_PORT & ~RW_MASK;    // RW = 0
        LCD_CTRL_PORT = LCD_CTRL_PORT & ~E_MASK;     //  E = 0
        large_delay(200);                            // 16ms delay

        LCD_DAT_PORT = 0x38;                 // set 8-bit mode
        pulse_E();
        large_delay(50);                     // 4.1ms delay

        LCD_DAT_PORT = 0x38;                 // set 8-bit mode
        pulse_E();
        large_delay(2);                      // 1.5ms delay

        LCD_DAT_PORT = 0x38;                 // set 8-bit mode
        pulse_E();
        large_delay(2);                      // 1.5ms delay

        lcd_cmd(0x06);                       // curser moves right
        lcd_cmd(0x01);                       // clear display
        lcd_cmd(0x0E);                       // display and curser on
}

#pragma OPTIMIZE (9)
//-------------------------------------------------------------------
// lcd_busy_wait - wait for the busy bit to drop
//-------------------------------------------------------------------
void lcd_busy_wait(void)
{
        LCD_DAT_PORT = 0xFF;
        LCD_CTRL_PORT = LCD_CTRL_PORT & ~RS_MASK;    // RS = 0
        LCD_CTRL_PORT = LCD_CTRL_PORT | RW_MASK;     // RW = 1
        small_delay(1);
        LCD_CTRL_PORT = LCD_CTRL_PORT | E_MASK;      //  E = 1
        do
        {
        // wait for busy flag to drop
                small_delay(1);
        } while ((LCD_DAT_PORT & 0x80) != 0);
}

//-------------------------------------------------------------------
// lcd_dat (putchar) - write a character to the lcd screen
```

```
//-----------------------------------------------------------------
char putchar(char dat)
{
        lcd_busy_wait();
        LCD_CTRL_PORT = LCD_CTRL_PORT | RS_MASK;     // RS = 1
        LCD_CTRL_PORT = LCD_CTRL_PORT & ~RW_MASK;    // RW = 0
        LCD_DAT_PORT = dat;
        pulse_E();
        return 1;
}

//-----------------------------------------------------------------
// lcd_cmd
//-----------------------------------------------------------------
void lcd_cmd(char cmd)
{
        lcd_busy_wait();
        LCD_CTRL_PORT = LCD_CTRL_PORT & ~RS_MASK;    // RS = 0
        LCD_CTRL_PORT = LCD_CTRL_PORT & ~RW_MASK;    // RW = 0
        LCD_DAT_PORT = cmd;
        pulse_E();
}

//-----------------------------------------------------------------
// lcd_clear - clears the display in the lcd screen
//-----------------------------------------------------------------
void lcd_clear(void)
{
        lcd_cmd(0x01); //-- clear LCD display
        lcd_cmd(0x80); //-- curser go to 0x00
}

void lcd_cursor(bit on)         // 1 displays curser, 0 hides it
{
        if (on)
                lcd_cmd(0x0E);
        else
                lcd_cmd(0x0C);
}

//-----------------------------------------------------------------
// Delay Routines
//-----------------------------------------------------------------
void small_delay(char d)
{
        while (d--);
}

void large_delay(char d)
{
        while (d--)
                small_delay(255);
}

void huge_delay(char d)
{
        while (d--)
                large_delay(255);
}
```

8

Embedded Microcontroller Based Fireworks Detonation System

8.1 Introduction

This chapter describes the implementation of a microcontroller based remote firing module to detonate fireworks. The system uses software running on a personal computer (PC) to control this remote module.

Traditional systems use very long runs of cable, up to several hundred meters, for each firework shell which is connected. This significantly increases the setup time and cost. If shorter wires are used, it will lead to placing technicians too close to the firework shells when they are fired. Such trade-offs between safety and setup cost/time should not be needed.

Because fireworks are classified as dangerous explosives the safety of all technicians and the public in the vicinity of the fireworks is of utmost importance. Safety mechanisms, both on the firing module and in the PC control software, must be implemented in order to avoid the unintentional detonation of any firework shell.

Large public displays, by their very nature, attract very large numbers of people. With such large crowds of people, all in anticipation of the fireworks display event, things can become stressful, and the possibility of human error can come into play; these safety features can be crucial to a successful outcome.

Large public fireworks displays must be choreographed to a high standard in order to meet public expectations. The designer of such a complex display requires the use of a computer controlled system in order to achieve the demanded level of timing and accuracy. But computer control of firework detonation alone is not enough; in order to reduce the complexity and workload of wiring to each electronic-match, a wireless system can be used to improve reliability, reduce setup time and cost. It will also improve the flexibility of positioning the control system, leading to an increase in the safety of pyro-technicians.

G. Sen Gupta and S.C. Mukhopadhyay: Embedded Microcontroller Interfacing, LNEE 65, pp. 157–198.
springerlink.com

Such wireless detonation and control systems are already commercially available but range in price from a few thousand dollars for a very basic system up to several thousand dollars for a high end system. Most of the systems, in the higher price brackets at least, include the ability to script firework detonations. There are only two software systems on the market that allow you to synchronize the detonations to music and simulate the show as a 3D graphical visualization. A system of this nature could cost the user upwards of $10,000 USD for the hardware and software. This is out of the reach of smaller national and regional firework display operators. A system is required that offers comparable features at a much reduced cost.

The designed system would be used for entertainment but with utmost safety. The remote firing module will be placed in close proximity to live explosives, and thus should be of rugged design. The remote module should be able to operate reliably at long ranges (up to 2 km) irrespective of weather conditions.

The overview of the designed system is shown in figure 8.1. The fireworks will be connected to the remote firing module which is connected to the control computer via wireless communication. There may be more than one remote firing module depending on the size of the fireworks display system.

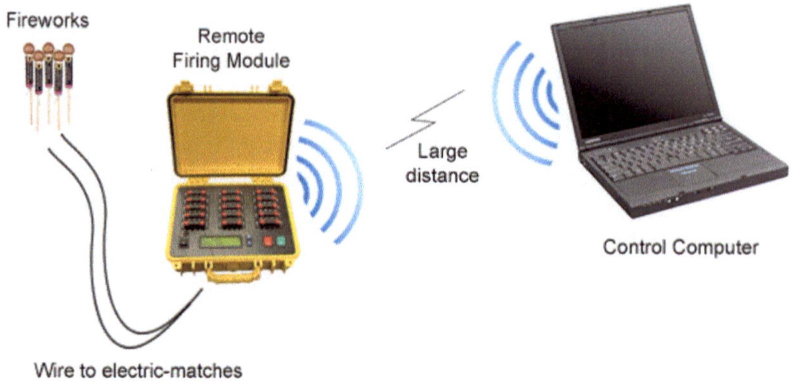

Fig. 8.1 System overview

8.2 Preliminary Version of the System

A 10-channel detonator board, using relays to switch the electric-matches, was initially designed as shown in figure 8.2. This detonator board needed to be connected to the Silicon Laboratories microcontroller development boards to do the processing and to interface with the MaxStream RF modems over RS232. There was no LCD display or ability to give diagnostic output from the firing module.

Fig. 8.2 Initial design of a 10-channel detonator board

The PC control software for the initial design was programmed in Visual Basic 6.0 and had a very simple graphical user interface (GUI). It also featured a basic scripting system whereby test and fire commands could be issued by the software with defined delay periods as shown in figure 8.3. The initial system had several limitations and there was a need to design a new system.

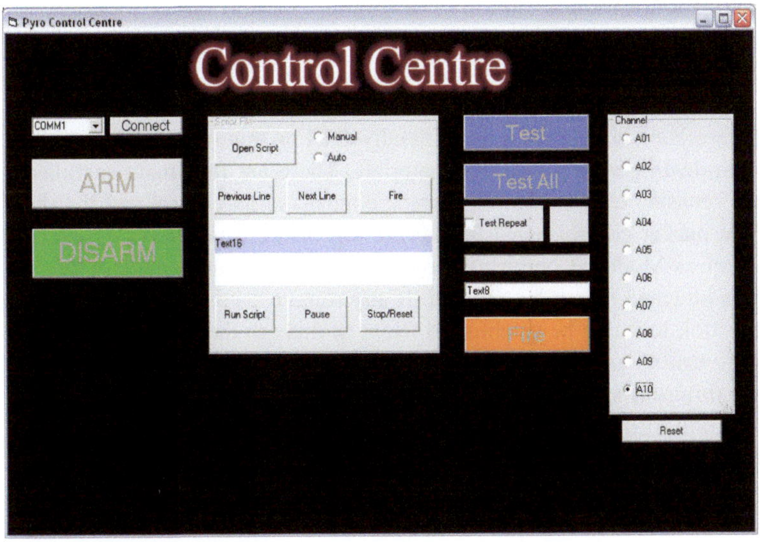

Fig. 8.3 Control software

8.3 Requirements

The proposed system should be composed of a remote firing module using solid-state devices as the switching components, and integrate the microcontroller board with the detonator board, so it is one complete, compact system. The module should provide some means to test all channels locally by a technician, and display the results on an LCD module. Safety switches are also to be incorporated into the device. The MaxStream RF module has been used as the wireless communications device for this application.

At least 25 controllable channels should be available to provide a good balance between price and available outputs. The channels should use some sort of reusable, easy to use connection to the module such as spring-loaded terminals or screw in terminals.

The PC control software should have a much more user-friendly GUI than the previous version and, if possible, the available commercial systems. It should allow both the manual testing and firing of electric-matches as well as automated firing by way of a scripting system.

The entire system should not be extraordinarily expensive as the objective of this project is to produce a system that could be commercialized and compete with existing products, targeted at smaller fireworks display operators who cannot afford overpriced commercial products.

8.4 Design and Implementation

8.4.1 Overview of Control Software

The PC Control Software was chosen to be written in C# as this is a modern development language with a lot of useful libraries and platform support. The two most important aspects that the software needed was the ability to control electric-matches both manually and through a scripting language.

The software created features a user-friendly GUI as this is very important especially when dealing with dangerous materials. There are four main sections to the software - master session control, manual control, script control and the show designer. The master control panel lays at the top of the user interface, and features the Wireless Module Connection box, where the serial port number that the RF modem is attached to can be selected. If any remote modules are powered on then a connection to it can be established. Next to this are the buttons for arming and disarming testing and firing. Testing and firing have separate arming buttons so as to keep things as safe as possible, thereby avoiding any possibility of a technician accidentally firing a channel when they only wanted to test it.

An XML configuration file contains various parameters that may be changed in future as deemed necessary, such as what resistances should constitute a short or an overloaded connection.

8.4.2 Manual Interface

The manual control interface allows the operator to view which modules are active, how many channels each contains and the status of each channel. From here channels can be individually tested and fired. Before testing or firing is allowed, the Arm Testing and Arm Firing buttons must be clicked to allow access to these features. This enables the onscreen controls, and sends a signal to the remote firing module indicating that firing and/or testing has been enabled. The firmware on the remote module keeps track of these states, and only allows firing or testing when both the PC and the module's safety switches are off. The manual interface screen is shown in figure 8.4.

Usually the resistance of an electric match is close to 2 ohms. The resistance of the electric match is measured to test whether the match is good, open or short-circuited. If the tested channel resistance is below 1.2 ohms the test result is classified as short and is displayed with a red background in the listview control. If it is between 1.2 and 2.5 ohms it is said to be a good connection and given a green background. If it is above this it is called a bad (overloaded) connection, and again displayed with a red background. If it is open-circuit it is displayed as such and given an orange background; this is because it is unknown to the program if the

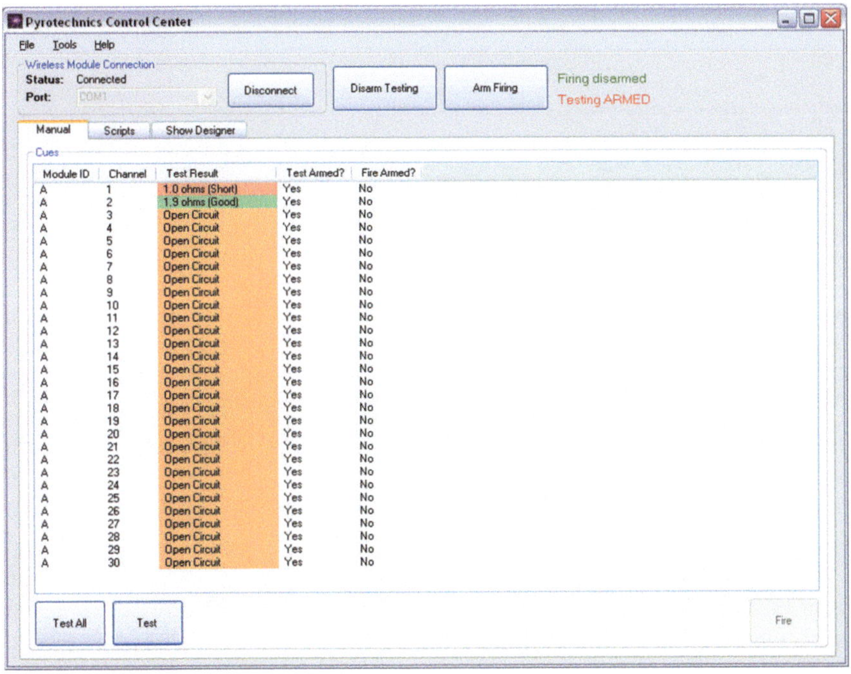

Fig. 8.4 Manual control interface

channel was left unconnected intentionally or not. These easy to understand colors give a quick indication as of the status of each channel, which is beneficial to the technicians.

A serial port communications log, as shown in figure 8.5, is also available from the drop down tools menu. This window displays all sent and received characters, which was extremely helpful during debugging. This can optionally be written to a text file.

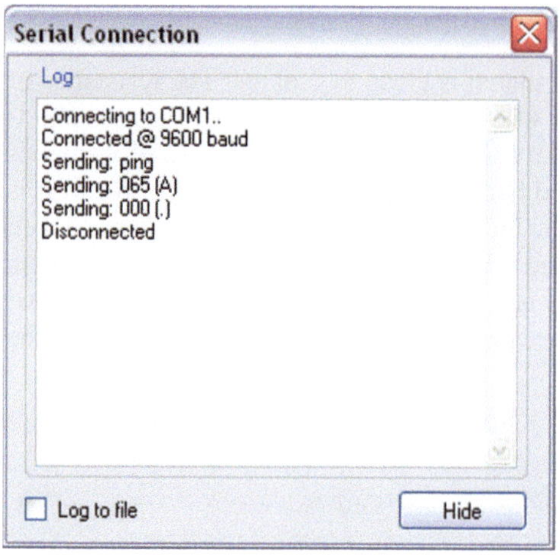

Fig. 8.5 Connection log window

8.4.3 Scripting Interface

The scripting interface is probably the most important section of the software, as it allows the execution of a completely automated fireworks display. The scripting interface window is shown in figure 8.6.

The module sections begin with the keyword "module" followed by the identifier of the module being referred to. A list of timecodes is then defined between the open and close braces. It is really necessary to have at what time each channel of the module should be fired, as all testing is accomplished through the manual interface. Therefore each line begins with the keyword "fire", then the channel number (these can be in any order), then the time from the start of script execution to fire each channel, in the format minutes ':' seconds ':' milliseconds. Comment lines begin with the '#' hash character.

A script text file must first be loaded into the software with the Browse button, even if it is only blank. It can then be viewed, edited and saved in a separate text-box pop-up window. Once the script is ready, the user clicks the "Setup" button.

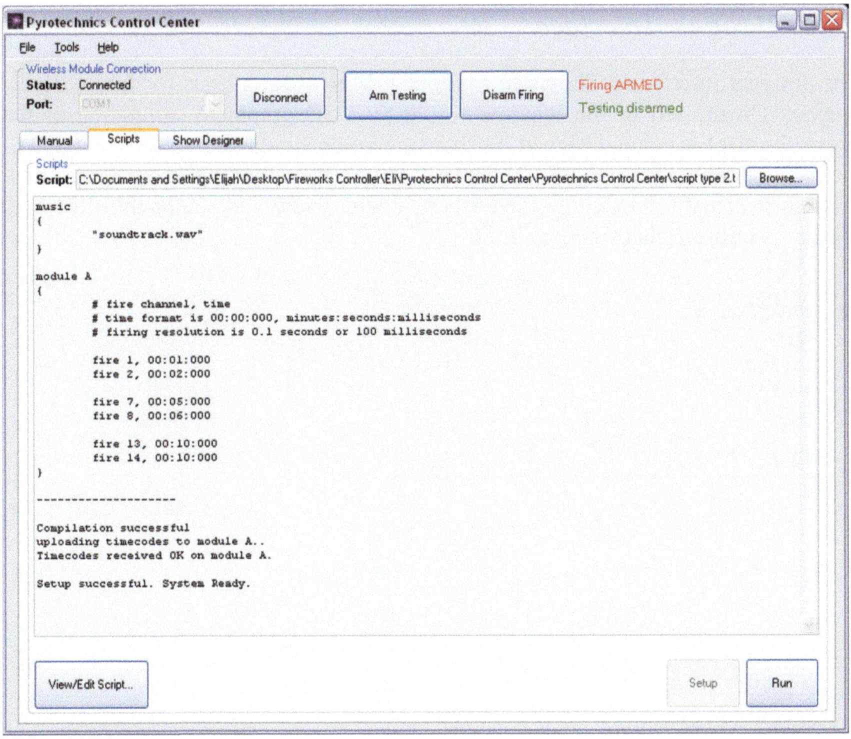

Fig. 8.6 Scripting control interface

This will compile the script and check for any syntactic errors, and validate what the user has entered; any errors are displayed in the output window, including what line they occur on. Setup will also fail if firing is not armed locally in the software or on any of the remote modules.

When this is okay the list of timecodes will be transmitted to the remote module. The remote modules will then send back an acknowledgement if the list of timecodes was received in good order. When setup is complete, the show can begin by clicking the "Run" button. This will transmit to the remote modules the go-ahead to start processing timecodes and begin firing.

If the user needs to cancel the display the "Stop" button can be pressed to send the cancel command to the remote modules.

It is a common practice with almost all of the existing commercial systems to use this concept of uploading timecodes to the firing modules for the reasons mentioned above. Some also allow the option of not needing to have a PC to initiate the firing process, in case the show technicians are not computer literate, though this concept is largely outdated in this day and age.

8.4.4 Designer Interface

The designer interface is shown in figure 8.7. It provides a basis for future development. On the left of the window would be a 3D graphical simulation of the fireworks display being designed. Such a simulation would be hugely beneficial to the creation and choreography of displays, letting the coordinators view the show prior to it actually happening, and letting clients view and sculpt the fireworks display to best suit their imaginations.

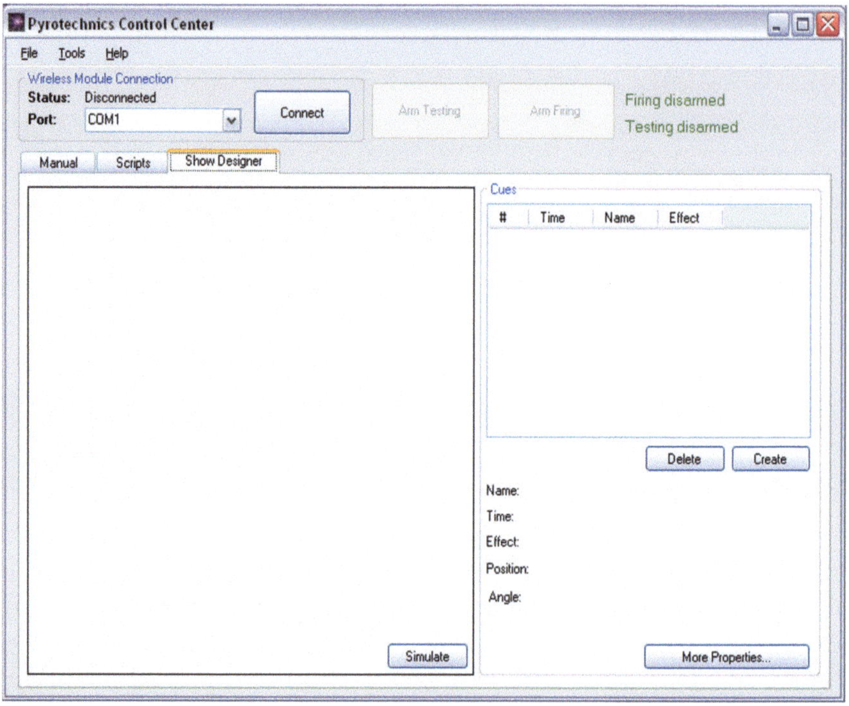

Fig. 8.7 Designer interface

8.5 Remote Firing Module

8.5.1 Overview and Methodology

The remote firing module is responsible for communicating with the control computer, displaying diagnostic results, and firing/testing electric-match connections. Each electric-match that can be fired by the module is connected to a "channel" (an individual, controllable electric-match connection). MOSFETs are used to switch current flow through the channels and these are controlled directly by an onboard microcontroller.

A large number of channels should be available on the module, as this will provide an adequate number of firework connections and reduce the number of these modules required for a large show. 30 channels have been built into the module. The system contains an LCD module for displaying status and diagnostic information, as well as local safety on/off switches for both firing and testing functionality for enhanced safety measures, again to protect the technicians wiring the system. Initially a small circuit board with 6 channels (including firing and testing current limit resistors, switching MOSFETs and zener diode for ADC readings) was designed and built in order to verify that the designed circuit for firing and testing would work. This helped to develop a large portion of the firmware for the system using the Silicon Labs microcontroller development boards. 1.8 ohm resistors were used in place of electric-matches for initial development.

After the successful design and testing of the experimental model, the full circuit was designed and fabricated. Solidworks was used for the CAD design of the control panel layout for all the mechanical components selected, and was machined on the CNC mill. The circuit boards, battery and panel were fitted inside a rugged, weatherproof plastic case.

When completed a final test was done using the actual electric-matches, with all going well. Testing and firing worked perfectly the first time.

8.5.2 Electric Matches

An electric match as shown in figure 8.8 is used to provide a small initial explosive charge to light a firework shell's primary fuse. They are typically constructed from lengths of 22-gauge insulated wire joined by a small bridge-wire coated in a pyrotechnic mixture that will ignite when heated. The electric-matches are from 1.5 to 2.5 ohms in resistance, and can be ignited by applying a current of approximately 1A (depending on model) through the match, with higher currents igniting the match at a faster rate. These are connected to the module using spring-loaded speaker terminals for ease of use.

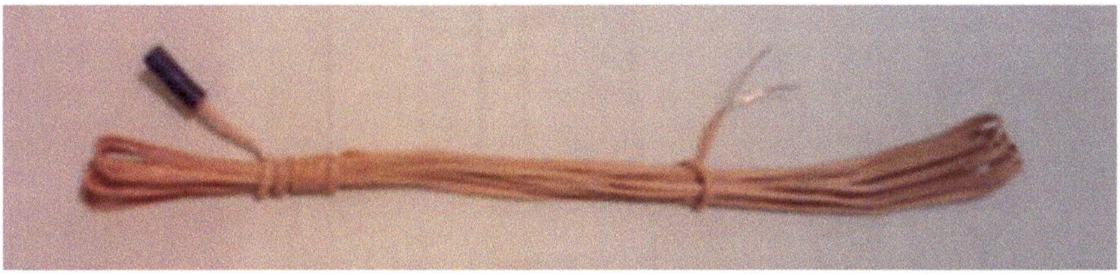

Fig. 8.8 An electric match used for detonating fireworks

Electric-matches have no-fire, and all-fire currents. Typically a no-fire current ranges from 50 to 100mA, but varies a lot between model and brand of match. This no-fire current means that all of the electric-matches are almost guaranteed to not fire/ignite at this current. This is very useful for determining the maximum current limit to allow when testing the electric-matches for continuity. Conversely an all-fire current dictates the minimum firing current required to successfully ignite all electric-matches, and typically ranges between 350 mA to 1 A.

8.5.3 User Interface

The user interface to the Remote Firing Module consists of two separate parts: a hardware panel with physical buttons, switches and connectors, and a graphical/textual display for dynamic content on the LCD.

In figure 8.9, the layout of the developed Remote Firing Module is shown. Spring loaded terminal connections numbering 1 through 30 are located at the top of the panel. Indicator LEDs for battery low and power are on the bottom left of the panel, next to the 2.5mm standard DC power jack for connection to a 16V+ un/regulated DC power supply for charging, with the power switch below this. The LCD is a 16 character by 2 line display, with large lettering and green illuminated backlight for ease of sight.

The small momentary push-button is the backlight control. The LCD backlight uses approximately 100mA of current when in use, and thus is a large waste of power. The Backlight button turns on or off the backlight. When the backlight is turned on, a timer will automatically turn it off after two minutes in order to conserve power. The technician simply needs to press the button in order to turn it back on again if they are working with the unit.

Fig. 8.9 Hardware user interface

The other momentary push-button is the test-all/diagnostic button. This button will immediately test all channels, and display the results locally by scrolling them across the bottom line of the LCD. This can aid technicians setting up the system, as they do not need to walk all the way back to the control station just to test if their wiring is good. The state of each channel, either "short", "open" or "OK" is displayed. More detailed results, i.e. the actual measured resistances across the terminals of each channel can only be viewed in the PC control software.

The fire arm and test arm safety switches are located on the bottom right of the module. These are large, illuminated, rocker switches. The switch must be in the on position to enable testing/firing locally on the module. When testing/firing is enabled both locally on the module and remotely in the PC control software, the switch will be illuminated. If firing/testing is enabled either locally or remotely, but not both, the switch will flash at a slow rate.

The top line of the LCD module displays several pieces of important information. The connection status- "Connected" or "Not Conn.". The module's ID at the far right, 'A' in figure 8.9, and also whether firing and testing are enabled, which is indicated by a 'T' for testing and a 'F' for firing. Testing/firing must be enabled both locally and remotely for this to be displayed, and indicates that all safety is off, and the channel connections are live. When scripted timecodes are active, "FIRE" is displayed in the bottom right corner of the LCD to indicate that firing on the module is in progress.

The bottom line displays the battery voltage in steps of 10%, which is updated in one minute intervals. The scrolling results of the local test-all button are also displayed here.

8.5.4 Operational Modes

There are three primary modes of system operation: normal, diagnostic, and charge.

Normal – the system by default boots into normal operating mode.

Diagnostic – this mode can be entered into by holding down the "test all/diagnostic" button when powering the system on. In this mode additional debugging information is displayed on screen in various states. This mode provides a very good basis for displaying additional debugging and development information in a simple manner.

Charge – charge mode is automatically entered into upon start up when a voltage reading of 16V or higher is detected on the input of the charger jack. In this mode all peripheral devices of the microcontroller are shut down, the LCD backlight is turned off, the RF module put into sleep mode and finally the microcontroller put into stop mode (disabling of the oscillator, effectively completely shutting down and disabling the chip). By doing this the whole system's current consumption is reduced to a few milliamps (the majority of this being the LCD display which requires 3mA to operate), and thus all of the current provided by the charging circuit can be used to charge the battery. By allowing the LCD to operate, the status of the system, "Charging", can be displayed, and the user made aware of this state.

To enable charging, first switch the module off and then connect the input voltage to the charger jack. The LCD will indicate that it has entered charge mode. Now switch the power switch on in order to connect the battery to the charger supply.

8.5.5 Wireless Network

The option chosen is to utilize industrial grade commercially available radio modems. The MaxStream XStream series of RF modems operate at 2.4 GHz and provide up to 16 km range with a high gain antenna, and 5 km with a dipole antenna. Communication with these devices is made simple, needing only a serial port on the PC, or can be talked to directly from a microcontroller that supports UART. One such module is shown in figure 8.10.

Fig. 8.10 XStream Modem Package

The devices support several types of communication topology: point-to-point, point-to-multipoint and broadcast. It is possible to configure each device with its own address ID, as -well as a receive-address-mask, so that it will only receive data from modules that match the required sender-address. The modules handle all necessary low level data protocols and encoding. They implement spread-spectrum frequency hopping to reduce the effects of noise and increase security. The modules come with PC software for easily configuring the devices.

8.5.6 Power Supply

A power supply is required that can deliver high currents (2 to 3A) while remaining small and portable. It must also last for several hours of continuous operation in the field.

Valve Regulated Lead-Acid (VRLA, also known as sealed lead-acid) batteries possess these traits, and are the only practical choice. The two main technologies available are gel-cell and Absorbed Glass Mat (AGM), both technologies are very similar and allow the battery to be stored, used and charged in virtually any position. These lead-acid batteries are reasonably compact, high capacity, and are capable of supplying large currents at 12V, making them ideal for our purposes.

Power consumption of the system is a major issue, as the device must be able to sit out in the field for up to 10 hours while a show is being setup. If all peripheral devices, including the LCD backlight consuming 100mA and the RF module consuming over 200mA, are active at all times the battery would not last long, therefore steps have been taken to reduce the on-time of these circuits.

Voltage regulation from 12V down to 3.3V at close to 60mA for the microcontroller, and at times 12V down to 5V at over 300mA means that a couple of Watts are dissipated by the small onboard voltage regulators, so the more current drawn by the system that can be reduced, the better.

The battery voltage is passed into 3.3V and 5V regulators to provide power to the microcontroller, LCD and RF module, the circuits are shown in figure 8.11. A diode is in place before the voltage regulators in order to provide protection to all sensitive components (microcontroller, capacitors, LCD, RF module etc.).

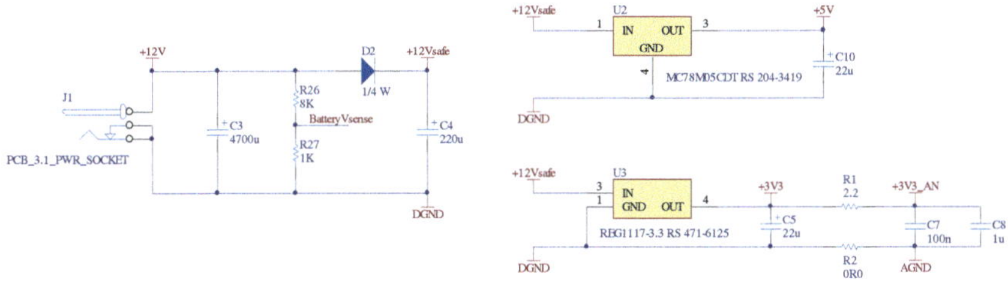

Fig. 8.11 System power supply circuit including voltage regulators

8.5.7 Battery Charger

A charging circuit for the self-contained VRLA battery has been implemented onboard in order to provide the simplest mechanism for recharging the battery. A DC power jack on the case allows for the input voltage to be connected.

VRLA batteries require a voltage across the terminals of 13.8V (2.3V per cell) in order to trickle charge. The battery will draw current it needs to recharge, though the initial current must be limited to 40% of C, the battery's Ah rating, for these types of battery in order to reduce the internal build up of gases; in this case 1.28A with the 3.2Ah battery used. The current drawn by the battery will slowly reduce as the battery reaches its full charge capacity.

LM317 (variable voltage regulator) has been used to suit the needs of this project, the circuit of which is shown in figure 8.12. It is a 13.8V regulator with 650mA current limit. This is a reasonably small current to draw from a supply, and at these high input voltages power (and thus heat) dissipation can be kept lower.

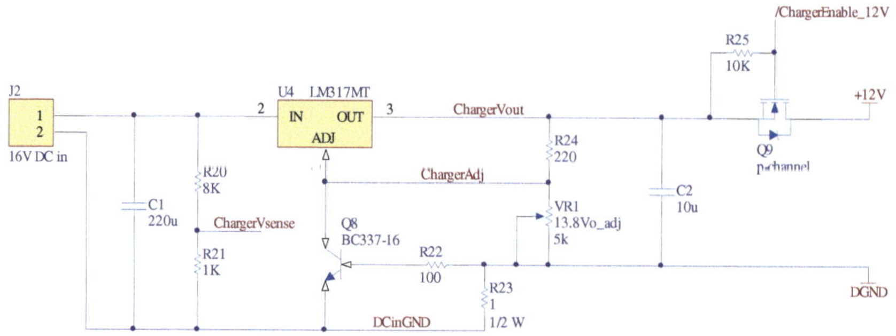

Fig. 8.12 VRLA charger

The input voltage is scaled down by approximately 1/8th in order to be fed into an ADC input of the microcontroller (this must be in the range 0 – 3.3V). The current drawn from the supply by the battery will return through the ground rail, passing through a 1 ohm, 0.5W resistor. When this current reaches approximately 650mA it will have a voltage of about 0.65V developed across it, thus turning on transistor Q8, pulling the GND/ADJ pin of the voltage regulator low, and cutting off the output. The output voltage of the regulator is given by:

$$Vout = 1.25V\left(1 + \frac{R2}{R1}\right)$$

Therefore with a value of 220 ohm for R1 (R24 in figure 8.12), a resistor of 2.2K (VR1 in figure 8.12) is required for R2 in order to get 13.8V at the output. As this is a non-standard value, a 5K trimpot has been used in order to tune the output voltage.

The battery voltage can be monitored by an ADC channel of the microcontroller, and the charging current enabled or disabled by use of a p-channel MOSFET on the positive rail.

Because the output required to charge the battery is 13.8V and the voltage regulator needs an approximate voltage drop of 2.2V in order to function, an input voltage of 16V DC (regulated or unregulated) or higher is required.

With 1 to 4 Watts being dissipated by the LM317 (depending on input voltage, 16-20V) at the maximum current of 650mA, a medium size heat sink has been affixed to it.

8.5.8 Firing and Testing

The requirement for firing is a device that can switch a high current through the electric-match in order to ignite it. While for testing it is required to allow a much smaller current through the match to determine whether the connection as well as the electric match is good or bad. This means it is possible to perform a simple continuity test, or measure the voltage developed across the connection to read the resistance, which would be a much more accurate measurement of the electric-match's state. This is because an electric-match may fail open circuit or short circuit, or the wires connecting it may not be completely connected or may short. In the case the connection is shorted, this cannot be detected with the simple continuity test, thus a large number of times the failure of the system cannot be detected. Therefore it is chosen to use the measurement of resistance as an input signal to ADC.

Relays are commonly used in similar situations where high current switching is required. They offer many benefits such as electrical isolation and ability to switch very large currents with minimal power loss, but also have several disadvantages, which primarily include needing relatively high currents to do the switching, large size and lifespan. Due to their mechanical nature, contact bounce and electrical arcing can reduce the lifespan of relays significantly.

By using power MOSFETs it is possible to switch quite high currents (up to 20A) with very small surface mount DPAK packages. These are also much cheaper per unit and increase the reliability of operation drastically. The circuit diagram used for firing and testing is shown in figure 8.13.

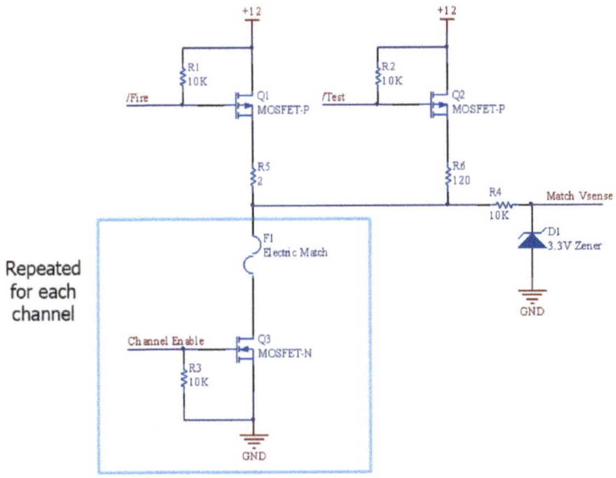

Fig. 8.13 Firing and testing circuit

To fire the electric-matches several amps of current are applied across the fuse-wire for a short burst of time. Each electric-match connection is controlled by a dedicated MOSFET with another common firing MOSFET to control current flow, this allows for several channels to be fired in parallel. Current is limited by a 2 ohm (R5) power resistor, capable of handling up to 50W power dissipation for 5 seconds; this in series with the electric-match will deliver about 3A of current. For testing the /Fire is made LOW and the /TEST is made HIGH, the channel corresponding to the target electric match is enabled.

8.6 Central Control Circuit

A Silicon Labs C8051F020 microcontroller is the heart of the module, providing all processing abilities, and interfacing with the RF module and LCD. This microcontroller is ideal for the needs of this project as it has 64 GPIO lines, onboard ADCs, UART and several timers. The details of the connections are shown in figure 8.14.

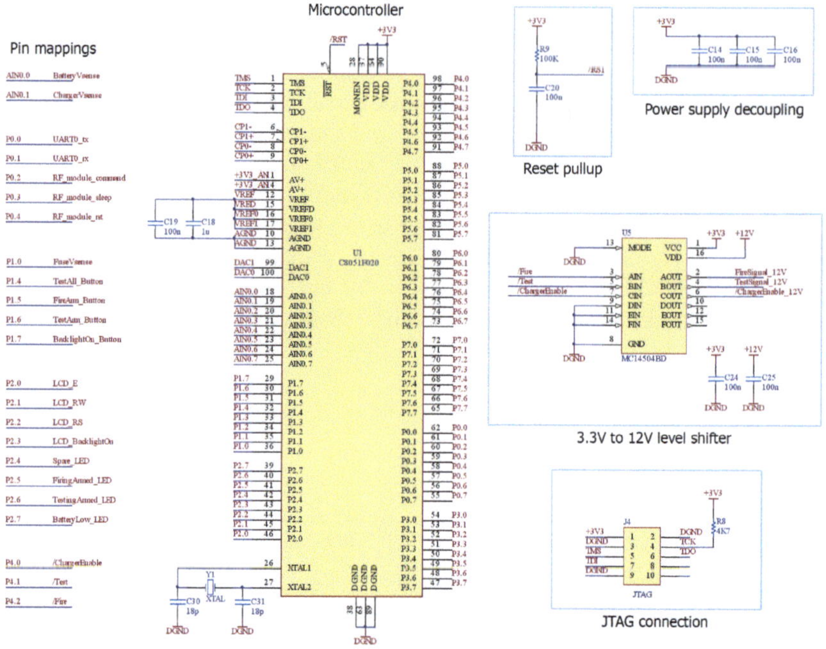

Fig. 8.14 Microcontroller schematic

The microcontroller is operating at a frequency of 22.118450MHz, as this divides down nicely for UART baud rates. Nearly every available IO resource of the microcontroller has been taken advantage of. Control and status pin mappings can be seen in figure 8.14. Along with P3 used for the LCD data lines, all of the pins of ports 5, 6 and 7 and some of ports 0 and 4 are used for the 30 channel control MOSFETs. The choice of which IO pins to use for what purpose was largely determined by their location on the physical device when routing the PCB.

8.6.1 LCD Control Circuit

LCD module MCC162B2-2 with 16 characters 2 lines has been used. The LCD module is connected by a ribbon cable to a 2x8 pin header on the system board, and uses all pins on port 3 for data lines, and 3 pins of port 2 for control lines. A 10 kohms trimpot provides contrast adjustment. The backlight of the LCD is an array of LEDs with a constant voltage drop of 4.2V. The current requirement for the backlight has been shown below.

$$100mA = \frac{5V - 4.2V}{8\Omega}$$

With such high power consumption it is not a good idea to have this running constantly, so a small n-channel MOSFET has been employed on the ground pin in order to allow the microcontroller to enable and disable the LCD backlight as required. The figures 8.15 and 8.16 show the necessary electronic connection diagram for the LCD module.

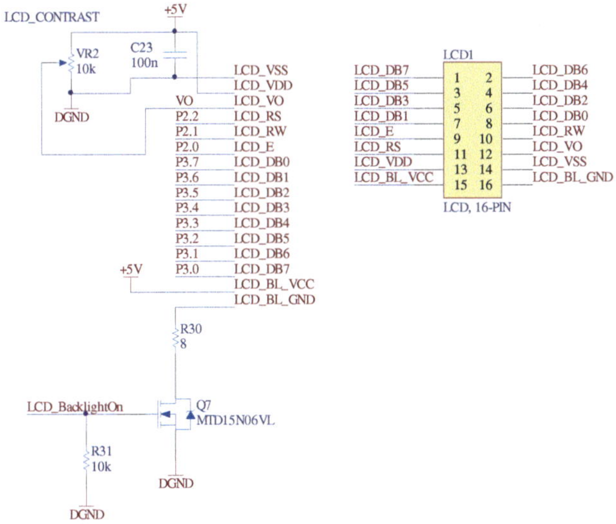

Fig. 8.15 LCD connection schematic

14	13	12	11	10	9	8	7	6	5	4	3	2	1	15	16
DB7	DB6	DB5	DB4	DB3	DB2	DB1	DB0	E	R/W	RS	VO	VDD	VSS	SLA	SLK

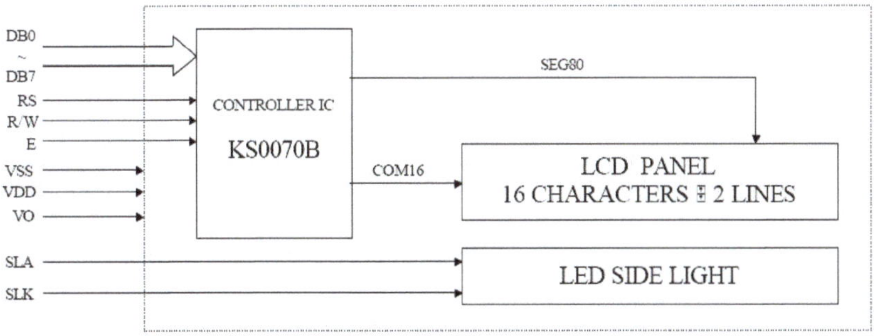

Fig. 8.16 LCD block diagram and pin out

8.6.2 *RF Modem Control Circuit*

The MaxStream RF modem is a small self contained package operating at 5V. The device can interface to 3.3V digital signals as it treats any voltage of 3.0V or higher on the inputs as a digital 'high'. Because of this fact, and the fact that the Silicon Labs microcontroller being used has 5V tolerant inputs, means we are able to directly connect the two devices without any special voltage level translation circuitry.

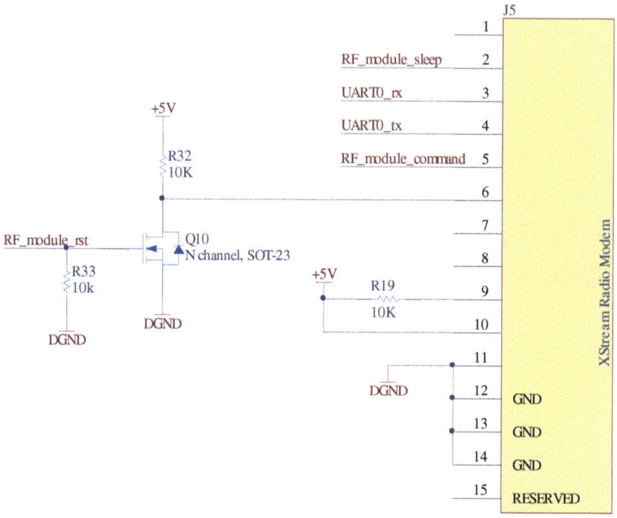

Fig. 8.17 RF module connection schematic

The device has an 11 pin header on the lower underside and a 4 pin header at the upper underside, which connect to sockets of the same dimensions on the system board. Many of the pins of the device are not needed for operation, only the TX and RX pins are required for use. In order to aid in reducing power consumption and ease of use connections to several other pins of the device have been made as is shown in figure 8.17.

The device needs to be configured in order to specify the function of many of the IO pins, as they often serve dual purposes. Configuration of the device is achieved in one of two ways: text mode, by entering command mode by sending "+++" and waiting for one second before issuing textual character commands; or through binary command mode, where commands can be issued by pulling the command pin high and then sending more efficient binary command bytes. Binary command mode must first be enabled through the text command mode, as the *CMD* pin defaults to another purpose.

This *CMD* pin has been connected to an IO line as executing binary commands is much more simple and far less time consuming than text commands. The *RST* pin has also been connected as a future-proof precaution through a small n-channel MOSFET so that the device can be reset by software if any problems with it occur.

The *sleep* pin has been connected in order to cut down the high current requirements of the modem, which uses 150mA when transmitting, and 80mA when receiving. By default the device is always idling in receive mode, and therefore needs to be told when it can enter into sleep mode. During sleep mode the device will consume less than 1 μA of current.

There are several sleep modes available, and the one to be used must be configured on the device.

- Pin-sleep – In this mode the modem stays asleep as long as the *sleep* pin is high. This mode is really useful to systems which transmit data only, as the device is not able to detect data being sent to it for it to receive while sleeping. Systems such as remote data-loggers transmitting at sporadic intervals would make best use of this sleep mode.
- Serial-port sleep – This mode is very similar to the pin-sleep mode, except that the device remains asleep until data is detected on the DI input pin. Therefore it remains sleeping until data is transmitted locally from the module; unfortunately it cannot detect when data is being sent to the device and wake during this mode.
- Cyclic-sleep – In cyclic-sleep mode the device enters sleep mode when there is no RF activity, i.e. when it is not transmitting or receiving. The modem will wake itself from sleep mode at user-configurable intervals (in the range 0.1 to 2 seconds) in order to check if data is being sent to the device. The pin-wakeup mode of the device must be enabled in combination with this mode to allow us to wake the device by de-asserting the *sleep* pin when we wish to transmit from the remote module. This is the most ideal sleep mode for our situation, and the one being used on the device.

8.6.3 IO Control Circuit

There are four LEDs connected to the system: one which is wired to 3.3V, which is the power on indicator, and three controlled by the microcontroller. A 2x5 pin header was used for these and with one pair of pins remaining it was decided to have a spare LED/IO line for use if future requirements demand it. This could prove quite useful as the hardware would not need to be extensively redesigned and rebuilt to incorporate only a small change.

Four inputs are connected to the microcontroller, two momentary push-buttons and two large rocker switches. These are connected with a 100K pull-up resistor and a 4.7K resistor into the microcontroller. The connected switch grounds the input line when in the on-state. Figures 8.18 and 8.19 show the details of the connections.

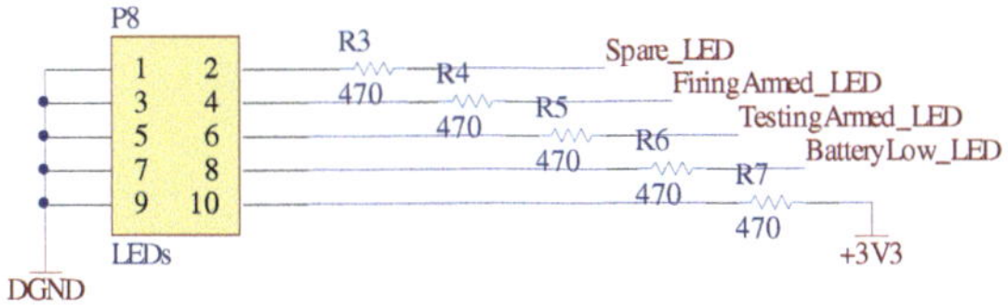

Fig. 8.18 LEDs schematic

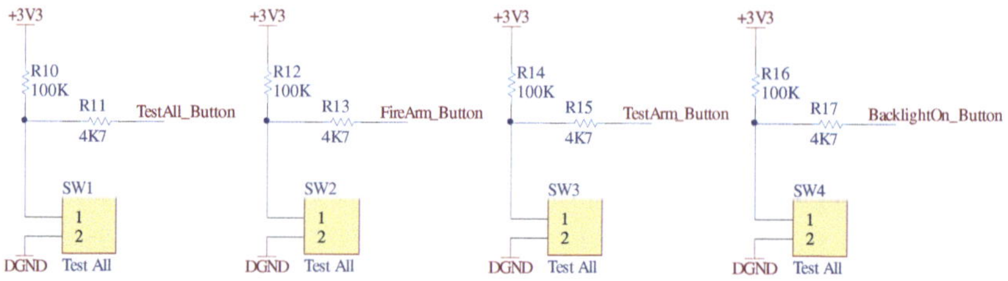

Fig. 8.19 Switches schematic

8.7 Developed Hardware

Some pictures of the developed system are shown in figures 8.20 to 8.25 which are self explanatory. The enclosure of the complete system, as is shown in figure 8.24, was chosen because it is made of very strong plastic, and has a waterproof rubber

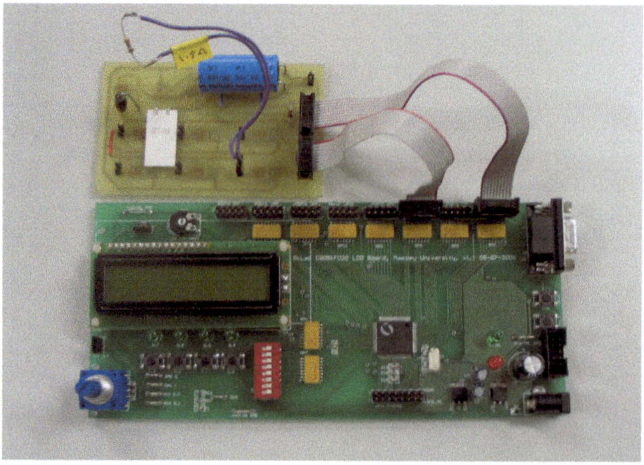

Fig. 8.20 Development system setup

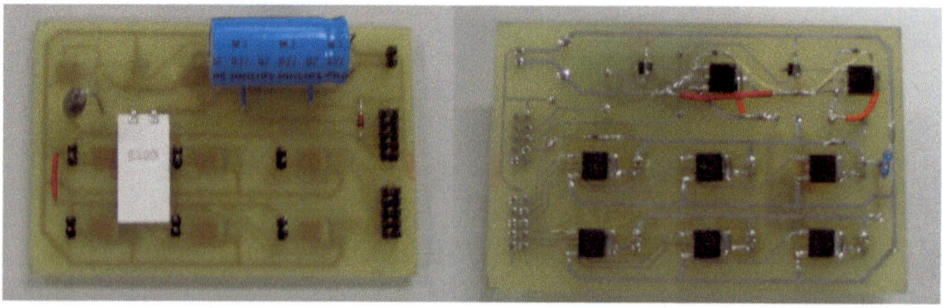

Fig. 8.21 Detonator board top/bottom

Fig. 8.22 Completed firing system board with attached RF module

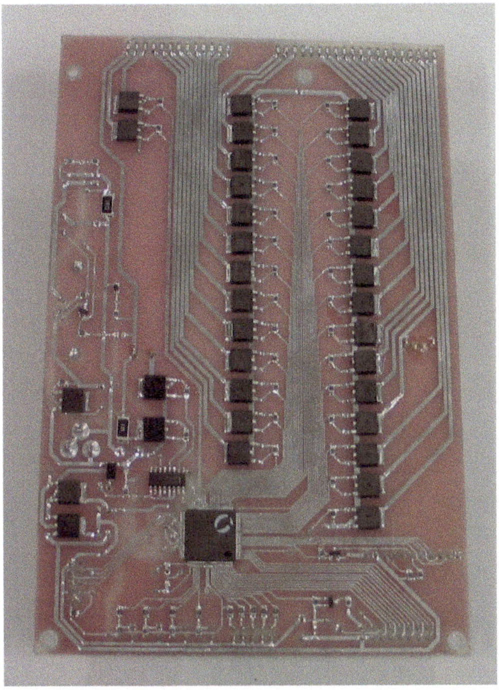

Fig. 8.23 Underside view of firing system board

Fig. 8.24 Enclosure and hardware control panel

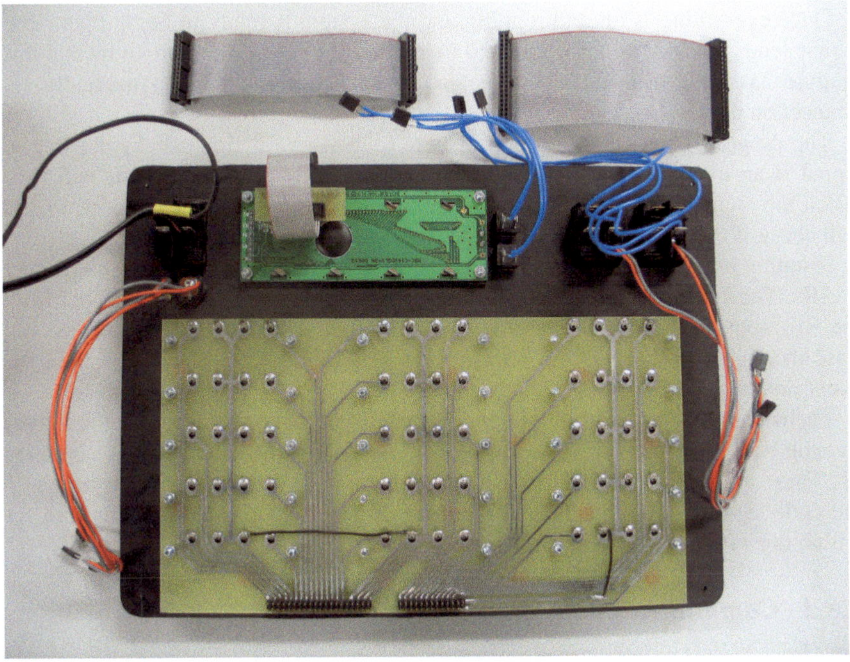

Fig. 8.25 Underside view of hardware control panel

seal. The panel was designed using Solidworks and machined on the CNC mill from a sheet of 3mm thick PVC. The case containing battery, controller board, LCD module and panel with spring-loaded terminal connections feels nice and heavy, of rugged design, as the system has been designed for use in a harsh outdoors environment controlling explosive materials.

The main system board is connected to another PCB on the underside of the top panel, to which the speaker terminals are soldered; this meant the connections would be a lot neater, eliminating the need for a large number of messy wire connections.

A small adapter PCB had to be made in order to connect the 1x16 pin LCD connection to the 2x8 ribbon cable header on the main system board.

8.8 Firmware

8.8.1 Overview

The microcontroller firmware is programmed in C, using the SiLabs IDE and the SDCC compiler. When the module powers on, it first of all initializes all variables to zero or their default values. After this it sets ups the microcontroller peripherals and subsystems before checking which operational mode to start in, normal, diagnostic or charge.

Once everything is up and running and the system is ready to begin operation. It first sends out the “hello” command to let the PC control program know that this module has been activated; this means the user doesn’t have to manually re-connect on the GUI.

The program then enters into an infinite loop where it first checks the flag to signal if an event timeout has occurred and needs to be processed. It will then check the command packet received flag, and process it if necessary. Otherwise it will check if the state of any of the input buttons or switches has changed.

Running in parallel to this are two interrupt service routines, UART0 and Timer 3 ISRs. The UART interrupt service routine will run when a byte is received from the RF module, it will process it and when the entire packet has been received raise the signal to the main loop to process it. Timer 3 serves as the basis of the event processing system.

Each channel’s port and IO pin are contained in lookup tables to be able to easily map between a channel number (i.e. 0 – 29) and the port and pin it is contained on. These lookup tables are stored in *xdata* memory. By doing this it has tidied up the code significantly, and IO pin numbers do not need to be hunted around for within the whole program to change when needed.

8.8.2 Communications

Every packet sent and received in the system begins with a four byte command string; this reduces the processing overhead required by the microcontroller. This is followed by the address of the module the packet is intended for, or has originated from. Any subsequent data specific to the command or response being sent is then appended to the packet. Table 8.1 provides the details of the packet structure.

Table 8.1 Packet structure

Index	Size	Type	Description
0	4	char	Command string ID, e.g. “badc”
4	1	char	Module ID, e.g. ‘A’, ‘B’, etc
5	1	int	Message data length
6	*	byte	Message data

The microcontroller’s UART0 peripheral device is used to accomplish the serial communication. These pins talk at standard 3.3V logic levels to the XStream radio modem, no +/-10V level conversion ICs are necessary. The XStream modules operate only at 9600 baud, so this is the bit rate used to configure UART0 and the reason the crystal frequency is at 22.118450 MHz, this frequency eliminates any timing errors for the UART baud rates.

When a command is to be received, the interrupt routine checks that the correct number of bytes (six for the packet header + the number specified for the message data length) has been received. The main program is then signalled, and the first four byte command string is decoded into a one byte integer identifier that is used to quickly check which command it is within the program. The command is then executed, or an error command is sent back to the PC if the command received could not be identified.

8.8.3 Command and Response Set

Commands are sent from the PC to the remote module, and are responsible for controlling the core actions of the module, such as testing and firing channels and requesting status responses from the module, including safety switch states, number of channels available on the module etc.

PC to Module available commands:

"**ping**" - ensure module is responsive, will respond with pong
"**fire**" - fire ematch on specified channel
"**test**" - test ematch on specified channel
"**tall**" - test all ematchs on specified module and send back results
"**fstc**" - set fire timecodes
"**tbeg**" - fire by timecodes begin
"**tcan**" - fire by timecodes stop
"**fdis**" - firing disable
"**fena**" - firing enable, ie. safety off
"**tdis**" - testing disable
"**tena**" - testing enable, ie. safety off
"**gtst**" - get test safety switch state
"**gfir**" - get fire safety switch state
"**gcnt**" - get addressable channel count of module

Module to PC available commands:

"**badc**" - bad command recv'd
"**pong**" - sent in response to a ping packet
"**busy**" - the module is busy processing the previous command
"**ftst**" - fire timecodes set okay
"**tres**" - test result, 2 byte message data is channel, then 8bit ADC value
"**tsws**" - test safety switch state, data is byte 0x00 for off, 0x11 for on
"**fsws**" - fire safety switch state, data is byte 0x00 for off, 0x11 for on
"**ccnt**" - channel count, message data is one byte containing the number of addressable channels on this module
"**helo**" - module has been powered on, broadcast to let PC control software know the module exists
"**eukn**" - error, an unknown error has occurred in the module
"**efdi**" - error, could not fire, firing disabled either locally or remotely
"**etdi**" - error, could not test, testing disabled either locally or remotely
"**erng**" - error, the requested channel to fire or test is out of range
"**etns**" - error, firing timecodes not set (when tbeg command issued)

8.8.4 Event System

The module's event processing system is the key to the module's operation. Every two milliseconds the timer 3 ISR (interrupt service routine) checks if there are any active event objects, and if so subtracts a value of 2 from its milliseconds remaining state variable. When this reaches zero the global *event_timeout* signal variable is set to 1, and the main program can then process it (i.e. call its associated event handler function and either remove the event, or set it back up to repeat).

Events are stored in an array of *event_t* object types. Where each event is given a unique ID with which it can be referred to. The *remain* variable keeps track of how many milliseconds remain until the event triggers, and the period variable contains the value that should be reloaded into the *remain* variable if the event is set to repeat. A callback function must also be specified that can be called when the event triggers to handle the event.

```
struct event_t
{
        BOOL repeat;

        unsigned char id;

        unsigned int remain; //-- in milliseconds
        unsigned int period;

        void (*callback)();
};
```

Events are started using the following function

```
void start_event(unsigned int milliseconds, BOOL repeat, void
                (*callback)(unsigned char), unsigned char id);
```

And they may be stopped at any time with

```
void stop_event(unsigned char id)
```

The timer 3 ISR runs at a clock frequency of 499.999 Hertz i.e. an interval of very, very close to 2 milliseconds. This frequency was chosen because the timer's reload value is a 16-bit integer (i.e. it must be in the range 0x0000 to 0xFFFF and cannot be a floating point number).

$$\text{Timer reload value} = 0\text{xFFFF} - \frac{\text{sysclock}}{\text{timerfreq}}$$

The timer reload value counts up to the max (0xFFFF), therefore the value we want to time for is subtracted from this.

$$\frac{sysclock}{timerfreq} = \frac{22114580}{500} = 44236.9 \approx 44237$$

A period of 500 Hz has been chosen, with a time period of 2 ms, because this value gave a result that was as close as possible to a full integer value compared to other millisecond periods. This in turn increases the timing accuracy significantly. The timer period also needed to be as small as possible to allow the highest resolution possible for events, and needed to be an integer number of milliseconds so event times could be accurately kept track of.

Therefore the real operational period of the timer is

$$\frac{\text{sysclock}}{\text{reloadvalue}} = \frac{22118450}{44237} = 499.998869\text{Hz}$$

This allows us to keep extremely accurate time with a resolution of two milliseconds on the remote firing module. This is accurate enough to keep the system in synchronization for the firing timecodes.

There are nine different events that are used on the module:

- Fire – keep output on for 500 ms in order to ensure matches are ignited successfully.
- Fire timecodes – repeated event every 100ms to keep time for firing on scripted timecodes when necessary.
- LCD Backlight – keeps the backlight on for two minutes at a time and then turns it off save power.
- Debounce – disables button input for 30ms when it is first detected in order to avoid multiple detections.
- Battery – reads battery voltage at one minute intervals to keep an accurate display of the battery's charge percentage.
- Scroll – scrolls the test-all display when the diagnostic button is pressed.
- Flash fire – flashes the fire armed LED when necessary.
- Flash test – flashes the test armed LED when necessary.
- Flash battery low – flashes the battery low warning LED when necessary.

In conclusion it can be said that a smart, compact and efficient fireworks detonation system based on embedded microcontroller has been designed and fabricated. The developed system has been tested satisfactorily in field trails.

A8 Chapter Appendix

Microcontroller Code Listing

```
/*
Purpose: Microcontroller firmware for Detonator test board for fire-
works control system project.

    Receives commands from control PC software over RS232.
    -------------------------------------------------------------
    PC-to-module packet layout:

    Index       size    type    description
    0           4       char    command string ID, eg "fire"
    4           1       char    ModuleID, eg. "A" or "B" etc
    5           1       int     channel

    Only "fire", "test", "chst" and "chun" commands use/send the
    channel byte. None of the strings are null terminated.

    Available commands:

    "ping" - ensure module is responsive, will respond with pong
    "fire" - fire ematch on specified channel
    "test" - test ematch on specified channel
    "tall" - test all ematchs on specified module; send back results.
    "fdis" - firing disable
    "fena" - firing enable, ie. safety off
    "tdis" - testing disable
    "tena" - testing enable, ie. safety off
    "gtst" - get test safety switch state
    "gfir" - get fire safety switch state
    "gcnt" - get addressable channel count of module
    -----------------------------------------------------------------
    module-to-PC packet layout:

    Index       size    type    description
    0           4       char    command string ID, eg "badc"
    4           1       char    ModuleID, eg. "A" or "B" etc
    5           *       byte    message data

    Available commands:

    "badc" - bad command recv'd
    "pong" - sent in reponse to a ping packet
    "busy" - the module is busy processing the previous command
    "tres" - test result, 2 byte message data is channel byte, then
             8bit ADC value
    "tsws" - test safety switch state, message data is byte 0x00 for
             off, 0x11 for on
    "fsws" - fire safety switch state, message data is byte 0x00 for
             off, 0x11 for on
    "ccnt" - channel count, message data is one byte containing the
             number of addressable channels on this module
    "helo" - module has been powered on, broadcast to let PC control
             software know the module exists
    "eukn" - error, an unknown error has occurred in the module
    "efdi" - error, could not fire, firing disabled either locally or
             remotely
    "etdi" - error, could not test, testing disabled either locally
             or remotely
```

```
    "erng" - error, the requested channel to fire or test on is out
            of range
    ----------------------------------------------------------------
    Fire armed switch at DIP switches #1
    Test armed switch at DIP switches #2
    Test All button at push-buttons #1
*/

#include <stdio.h>
#include <c8051f020.h>
#include "LCD.h"
#include "utils.h"

#define CMD_UNKNOWN                           0
#define CMD_PING                              1
#define CMD_FIRE                              2
#define CMD_TEST                              3
#define CMD_TESTALL                           4
#define CMD_FIRE_ENABLE                       5
#define CMD_FIRE_DISABLE
#define CMD_TEST_ENABLE                       7
#define CMD_TEST_DISABLE
#define CMD_GET_FIRE_SW                       9
#define CMD_GET_TEST_SW                       10
#define CMD_GET_CHANNEL_COUNT                 11
#define RESPONSE_BAD_COMMAND                  101
#define RESPONSE_BUSY                         102
#define RESPONSE_PONG                         103
#define RESPONSE_TEST_RESULT                  104
#define RESPONSE_TEST_SW_STATE                105
#define RESPONSE_FIRE_SW_STATE                106
#define RESPONSE_FIRING_DISABLED              107
#define RESPONSE_TESTING_DISABLED             108
#define RESPONSE_CHANNEL_OUT_OF_RANGE         109
#define RESPONSE_CHANNEL_COUNT                110
#define RESPONSE_MODULE_ACTIVE                111

#define MY_MODULE_ID   'A'     // MUST be different for each hardware
                               // module connected to the system
#define MY_CHANNEL_COUNT 6

#define fire_pin P1_7
#define test_pin P1_5
#define vsen_pin P1_3

// Types
struct recv_packet_type
{
    char command;
    char module;
    char channel;
};

// Pin mapping constants
char xdata channel_port_map[] =
{
    3, // 0
    3, // 1
    3, // 2
    3, // 3
    3, // 4
    3  // 5
};
```

```
char xdata channel_port_pinmask[] =
{
    0x08, // 0 at P3.3, i.e. 0000 1000
    0x01, // 1 at P3.0
    0x02, // 2 at P3.1
    0x40, // 3 at P3.6
    0x80, // 4 at P3.7
    0x20  // 5 at P3.5
};
// UART vars
volatile char recv_command_str[4];   // holds four character command
string before being decoded

volatile char recv_index;     // keeps track of which byte of the
                              // packet being received we are upto
volatile char packet_received;       // boolean flag

volatile struct recv_packet_type recv_packet;
// receiving packet buffer
volatile struct recv_packet_type proc_packet;
// processing packet buffer

volatile char tx_complete;

// state vars
bit busy;      // busy processing eg. local diagnostics, test all
bit connected; // connected with control PC, updated when cmd recvd
bit accepting_input;  // donot accept input during switch debouncing
bit fire_armed_local;
bit fire_armed_remote;
bit test_armed_local;
bit test_armed_remote;
bit switch_state_fire;
bit switch_state_test;
bit switch_state_testall;

// timing vars
unsigned int timer_remaining_ms;

void (*pfnEventCallback)(void);

// Prototypes
void do_command(struct recv_packet_type* pPacket);
char decode_command(char* str);
void update_display(void);

void send_response(char response_type);

void fire_channel(char channel);
void test_channel(char channel);

void test_all_channels(void);
void perform_local_diagnostics(void);
void start_fire_timeout(unsigned int ms);
void start_debounce_timeout(unsigned int ms);

void fire_timeout_callback(void);
void debounce_timeout_callback(void);

void start_event_timeout(unsigned int milliseconds, void
(*pfnCallback)(void));
```

```
void init_ADC1(void);
void init_UART0(void);

//-----------------------------------------------------------------
void init(void)
{
   WDTCN = 0xDE;       // disable watch dog timer
   WDTCN = 0xAD;

   OSCXCN = 0x67;                    // enable external oscillator
   while ((OSCXCN & 0x80) == 0);      // wait for xtal to stabilize

   OSCICN = 0x08;                    // disable internal oscillator

   //-- configure digital crossbar
   XBR0 = 0x04; // enable UART0 (which uses P0.0 and P0.1)
   XBR1 = 0x00;
   XBR2 = 0x40; // enable crossbar and weak pull-ups (globally)

   //-- configure ports (1 = push-pull output)
   P0MDOUT = 0x01;    // enable TX0 as a push-pull output
   P1MDOUT = 0xA0;    // enable P1.5 and P1.7 as push-pull output
                      // (test and fire pins)
   P2MDOUT = 0xFF;    // P2 all to push-pull
   P3MDOUT = 0xFF;    // P3 all to push-pull


   P74OUT = 0xC8;     // output configuration for P4-7
             // (P7[7:4] Push Pull) - fire/test status LEDs
             // (P7[0:3] Push Pull) - control lines for LCD
             // (P6 Open-Drain)- data lines for LCD
             // (P5[7:4] Push Pull) - 4 LEDs
       // (P5[3:0] Open Drain) - 4 push-button siwtches (input)
       // (P4 Open Drain) - 8 DIP switches (input)
   fire_pin = 0;
   test_pin = 0;

   P2 = 0x00;
   P3 = 0x00;

   P7 &= 0x0F;

   //-- write a logic 1 to those pins which are to be used for input
   P5 |= 0x0F; // development board 4 push-button switches
   P4 = 0xFF; // DIP switches

   //-- init peripherals
   init_UART0();
   init_ADC1();

   lcd_init();
   lcd_curser(0); // switch off curser
}

//-----------------------------------------------------------------
void init_ADC1(void)
{
   //-- setup pin P1.3 as ADC1.3 input
   REF0CN = 0x03;     // enable internal reference
   P1MDIN = 0xF7;     // P1.3 configured as analog input,
                      // others as output
   AMX1SL = 0x03;     // select AIN1.3 for input
   ADC1CF = 0xF5;     // highest possible conversion frequency with
                      // gain of 1
```

```
    ADC1CN = 0x80;          // enable ADC1 in continuous tracking mode,
                            // conversion initiated on write to AD1BUSY

    vsen_pin = 1;           // must explicitly write logic 1 to this open
                            // drain pin to be used for input (disable
                            // output driver)
}

//-----------------------------------------------------------------------
void init_UART0(void)
{
    //-- set up timer 1 to generate the baude rate (9600) for UART0
    CKCON |= 0x10; // T1M=1; timer 1 uses the sys clk 22.11845 MHz
    TMOD = 0x20;   // timer 1 in Mode 2 (8-bit auto-reload)
    TH1 = 0x70;    // baudrate = 9600
    TR1 = 1;       // start timer 1 (TCON.6 = 1)

    //-- set up the UART0
    PCON |= 0x80; // SMOD0=1 (UART0 baud rate divide-by-2 disabled)
    SCON0 = 0x50; // UART0 Mode 1, Logic level of stop bit ignored
                            // and Receive enabled

    //-- enable UART0 interrupt
    IE |= 0x10;
    IP |= 0x10;             // set to high priority level

    RI0 = 0; // clear the receive interrupt flag; ready to receive
//  TI0 = 0;
}

//-----------------------------------------------------------------------
void main(void)
{
    recv_index = 0;
    tx_complete = 1;

    packet_received = 0;

    busy = 0;
    connected = 0;
    accepting_input = 1;

    switch_state_fire = (P4 & 0x02) >> 1;
    switch_state_test = (P4 & 0x01);
    switch_state_testall = P5 & 0x01;

    test_armed_remote = 0;
    fire_armed_remote = 0;

    test_armed_local = switch_state_test;
    fire_armed_local = switch_state_fire;

    EA = 0;     // disable interrupts
    init();
    update_display();

    send_response(RESPONSE_MODULE_ACTIVE);
    EA = 1;     // enable interrupts

    // Loop forever
    while(1)
    {
      // Process received commands
      if(packet_received == 1)
```

```
    {
      busy = 1;

      do_command( &proc_packet );

      busy = 0;
      packet_received = 0;
    }

    // Process input
    if(accepting_input && !busy)
    {
      bit sw_test = (P4 & 0x01);
      bit sw_fire = (P4 & 0x02) >> 1;
      // shift concerned bit into position 1 so it is assigned to
       // the bit value type
      bit sw_diagnostic = (P5 & 0x01);

      // Check test armed switch
      if( switch_state_test != sw_test )
      {
              switch_state_test = sw_test;
              test_armed_local = switch_state_test;

              //start_debounce_timeout(30);

              send_response(RESPONSE_TEST_SW_STATE);
              send_byte(test_armed_local ? 0x11 : 0x00);

              update_display();
      }

      // Check fire armed switch
      if( switch_state_fire != sw_fire )
      {
              switch_state_fire = sw_fire;
              fire_armed_local = switch_state_fire;

              //start_debounce_timeout(30);

              send_response(RESPONSE_FIRE_SW_STATE);
              send_byte(fire_armed_local ? 0x11 : 0x00);

              update_display();
      }

      // Check local diagnostics button
      if( switch_state_testall != sw_diagnostic )
      {
              switch_state_testall = sw_diagnostic;

              if(switch_state_testall == 0)
              {
                      //start_debounce_timeout(4000);
                      perform_local_diagnostics();
              }
      }
    }
  }
}
```

```
//----------------------------------------------------------------
char decode_command(char* str)
{
   if(isequal(str, "ping", 4)) return CMD_PING;
   if(isequal(str, "fire", 4)) return CMD_FIRE;
   if(isequal(str, "test", 4)) return CMD_TEST;
   if(isequal(str, "tall", 4)) return CMD_TESTALL;
   if(isequal(str, "fena", 4)) return CMD_FIRE_ENABLE;
   if(isequal(str, "fdis", 4)) return CMD_FIRE_DISABLE;
   if(isequal(str, "tena", 4)) return CMD_TEST_ENABLE;
   if(isequal(str, "tdis", 4)) return CMD_TEST_DISABLE;
   if(isequal(str, "gfir", 4)) return CMD_GET_FIRE_SW;
   if(isequal(str, "gtst", 4)) return CMD_GET_TEST_SW;
   if(isequal(str, "gcnt", 4)) return CMD_GET_CHANNEL_COUNT;

   return CMD_UNKNOWN;
}


//----------------------------------------------------------------
void do_command(struct recv_packet_type *pPacket)
{
   switch(pPacket->command)
   {
   case CMD_PING:
     send_response(RESPONSE_PONG);
     break;

   case CMD_FIRE:

     // DEBUG
     P7 |= 0x80;
     delay_10ms(25);
     P7 &= ~0x80;
     delay_10ms(25);
     P7 |= 0x80;
     delay_10ms(25);
     P7 &= ~0x80;
     delay_10ms(25);
     P7 |= 0x80;
     delay_10ms(25);
     P7 &= ~0x80;
     delay_10ms(25);
     P7 |= 0x80;
     delay_10ms(25);
     P7 &= ~0x80;

     fire_channel(pPacket->channel);
     break;

   case CMD_TEST:

     // DEBUG
     P7 |= 0x20;
     delay_10ms(25);
     P7 &= ~0x20;
     delay_10ms(25);
     P7 |= 0x20;
     delay_10ms(25);
     P7 &= ~0x20;
     delay_10ms(25);
     P7 |= 0x20;
     delay_10ms(25);
     P7 &= ~0x20;
```

```
        delay_10ms(25);
        P7 |= 0x20;
        delay_10ms(25);
        P7 &= ~0x20;

        test_channel(pPacket->channel);
        break;

      case CMD_TESTALL:

        // DEBUG
        P7 |= 0x20;
        delay_10ms(25);
        P7 &= ~0x20;
        delay_10ms(25);
        P7 |= 0x20;
        delay_10ms(25);
        P7 &= ~0x20;
        delay_10ms(25);
        P7 |= 0x20;
        delay_10ms(25);
        P7 &= ~0x20;
        delay_10ms(25);
        P7 |= 0x20;
        delay_10ms(25);
        P7 &= ~0x20;

        test_all_channels();
        break;

      case CMD_FIRE_ENABLE:
        fire_armed_remote = 1;
        update_display();
        break;

      case CMD_TEST_ENABLE:
        test_armed_remote = 1;
        update_display();
        break;

      case CMD_FIRE_DISABLE:
        fire_armed_remote = 0;
        update_display();
        break;

      case CMD_TEST_DISABLE:
        test_armed_remote = 0;
        update_display();
        break;

      case CMD_GET_FIRE_SW:
        send_response(RESPONSE_FIRE_SW_STATE);
        send_byte(fire_armed_local ? 0x11 : 0x00);
        break;

      case CMD_GET_TEST_SW:
        send_response(RESPONSE_TEST_SW_STATE);
        send_byte(test_armed_local ? 0x11 : 0x00);
        break;

      case CMD_GET_CHANNEL_COUNT:
        send_response(RESPONSE_CHANNEL_COUNT);
        send_byte(MY_CHANNEL_COUNT);
        break;
```

```
    default: // CMD_UNKNOWN
      {
        send_response(RESPONSE_BAD_COMMAND);
      }
    }

    // update display
    if(!connected)
    {
      connected = 1;
      update_display();
    }
}

//-------------------------------------------------------------
void update_display(void)
{
    char write;

    // Test armed display
    if(test_armed_local && test_armed_remote)
      write = 'T';
    else write = ' ';

    lcd_goto(12);
    putchar(write);

    // Fire armed display
    if(fire_armed_local && fire_armed_remote)
      write = 'F';
    else write = ' ';

    lcd_goto(13);
    putchar(write);

    // Module ID
    lcd_goto(15);
    putchar(MY_MODULE_ID);

    // Connection state display
    lcd_goto(0);

    if(connected == 1)
      printf("Connected");
    else
      printf("Not Conn.");
}

//-------------------------------------------------------------
void fire_channel(char channel)
{
    // Ensure this is legal
    if(!fire_armed_local || !fire_armed_remote)
    {
      send_response(RESPONSE_FIRING_DISABLED);
      return;
    }

    if(channel >= MY_CHANNEL_COUNT)
    {
      send_response(RESPONSE_CHANNEL_OUT_OF_RANGE);
      return;
    }
```

```
    // Enable port pin
    port_or(channel_port_map[channel],
                        channel_port_pinmask[channel]);

    // Enable fire pin
    fire_pin = 1;

    start_fire_timeout(1000);

    /* taken care of in timeout callback
    delay_10ms(6);
    fire_pin = 0;
    port_and(channel_port_map[channel],
                        ~channel_port_pinmask[channel]);
    */
}

//-----------------------------------------------------------------------
void test_channel(char channel)
{
    char result;

    // Ensure this is legal
    if (!test_armed_local || !test_armed_remote)
    {
      send_response(RESPONSE_TESTING_DISABLED);
      return;
    }

    if (channel >= MY_CHANNEL_COUNT)
    {
      send_response(RESPONSE_CHANNEL_OUT_OF_RANGE);
      return;
    }

    EA = 0;     // we will want to disable interrupts while we do this
                // to ensure there is no chance of it taking longer
                // than it should and detonating something.

    // Enable output pins
    port_or(channel_port_map[channel],
                         channel_port_pinmask[channel]);

    test_pin = 1;

    // Give mosfets time to switch on
    delay_10us(5);

    // Get result
    ADC1CN &= ~0x20;   // clear interrupt flag, AD1INT = 0
    ADC1CN |= 0x10;    // start conversion, AD1BUSY = 1
    while( (ADC1CN & 0x20) == 0 );    // poll for measurement
                                      // complete, AD1INT = 1
    ADC1CN &= ~0x20;   // clear interrupt flag, AD1INT = 0

    result = ADC1;

    // Turn off test MOSFET
    test_pin = 0;

    // Give time for test to turn off and ground output through
    // channel mosfet (otherwise the fuse_positive track is left
    // floating)
    delay_10us(6);
```

```
    // Turn off channel MOSFET
    port_and(channel_port_map[channel],
                                ~channel_port_pinmask[channel]);

    EA = 1;

    send_response(RESPONSE_TEST_RESULT);
    send_byte(channel);
    send_byte(result);
}

//---------------------------------------------------------------
void perform_local_diagnostics(void)
{
    bit prev_armed_local = test_armed_local;
    bit prev_armed_remote = test_armed_remote;

    if (!busy)
    {
      busy = 1;

      test_armed_local = 1;
      test_armed_remote = 1;

      test_all_channels();

      test_armed_local = prev_armed_local;
      test_armed_remote = prev_armed_remote;

      busy = 0;
    }
}

//---------------------------------------------------------------
void test_all_channels(void)
{
    char i;

    // Ensure this is legal
    if (!test_armed_local || !test_armed_remote)
    {
      send_response(RESPONSE_TESTING_DISABLED);
      return;
    }

    // Perform test
    for(i = 0; i < MY_CHANNEL_COUNT; i++)
      test_channel(i);
}

//---------------------------------------------------------------
void send_response(char response_type)
{
    char* text = "eukn";

    // set response code string
    if (response_type == RESPONSE_BAD_COMMAND)
    text = "badc";
    else if (response_type == RESPONSE_PONG)
    text = "pong";
    else if (response_type == RESPONSE_BUSY)
    text = "busy";
    else if (response_type == RESPONSE_TEST_RESULT) text = "tres";
    else if (response_type == RESPONSE_TEST_SW_STATE) text = "tsws";
```

```
    else if (response_type == RESPONSE_FIRE_SW_STATE) text = "fsws";
    else if (response_type == RESPONSE_CHANNEL_COUNT) text = "ccnt";
    else if(response_type == RESPONSE_FIRING_DISABLED) text = "efdi";
    else if(response_type == RESPONSE_TESTING_DISABLED) text ="etdi";
    else if(response_type == RESPONSE_CHANNEL_OUT_OF_RANGE)
                                        text= "erng";
    else if(response_type == RESPONSE_MODULE_ACTIVE)text = "helo";

    // send data
    send_byte(text[0]);
    send_byte(text[1]);
    send_byte(text[2]);
    send_byte(text[3]);
    send_byte(MY_MODULE_ID);
}

//---------------------------------------------------------------------
void start_fire_timeout(unsigned int ms)
{
    start_event_timeout(ms, fire_timeout_callback);

    P7 |= 0x80;
}

//---------------------------------------------------------------------
void start_debounce_timeout(unsigned int ms)
{
    accepting_input = 0;

    start_event_timeout(ms, debounce_timeout_callback);
}

//---------------------------------------------------------------------
void fire_timeout_callback(void)
{
    int i;

    P7 &= ~0x80;

    fire_pin = 0;

    // Turn off all channels
    for(i = 0; i < MY_CHANNEL_COUNT; i++)
      port_and(channel_port_map[i], ~channel_port_pinmask[i]);
}

//---------------------------------------------------------------------
void debounce_timeout_callback(void)
{
    accepting_input = 1;
}

//---------------------------------------------------------------------
void start_event_timeout(unsigned int milliseconds, void
(*pfnCallback)(void))
{
    TMR3CN = 0x00; // stop and clear, use sysclock/12


    pfnEventCallback = pfnCallback;
    timer_remaining_ms = milliseconds;
```

```
    TMR3RL = 0x0000;
    TMR3 = 0xFFFF;  // reload immediately

    EIE2   |= 0x01; // enable interrupts for timer 3
    TMR3CN |= 0x04; // start timer 3
}

//-----------------------------------------------------------------------
void UART0_ISR(void) interrupt 4
{
    /* TODO
       Start timer so if a whole valid packet is not recved withen X
       Milliseconds it sends a recv-timeout response to the PC and re
       sets recv_index.

       This is to safe-guard against possible transmission errors and
       something getting out of sync.
    */

    int packet_length;
    char received_byte;

    P2_1 = !P2_1;

    // interrupt caused by received byte
    if (RI0 == 1)
    {
       received_byte = SBUF0;  // read the input buffer
       RI0 = 0;                // clear the interrupt flag

       recv_index++;

       // determine packet length
       packet_length = 5;

       if ( (recv_packet.command == CMD_FIRE) ||
         (recv_packet.command == CMD_TEST) )
       {
         packet_length = 6;
       }

       // get command str
       if (recv_index <= 4)
       {
         recv_command_str[ recv_index-1 ] = received_byte;

         if (recv_index == 4)
            recv_packet.command = decode_command(recv_command_str );
       }

       // get module id
       if (recv_index == 5)
       {
         recv_packet.module = received_byte;
       }

       // get channel id
       if (recv_index == 6)
       {
         recv_packet.channel = received_byte;
       }

       // done
       if (recv_index == packet_length)
```

```
        {
          recv_index = 0;

          if ( recv_packet.module == MY_MODULE_ID )
          {
                 // do not process if busy processing previous command
                 if (packet_received || busy)
                 {
                        send_response(RESPONSE_BUSY);
                 }
                 // signal program
                 else
                 {
                        proc_packet.module     = recv_packet.module;
                        proc_packet.channel    = recv_packet.channel;
                        proc_packet.command    = recv_packet.command;

                        packet_received = 1;
                 }
          }

          recv_packet.module = 0;
          recv_packet.command = CMD_UNKNOWN;
        }
      }

      // interrupt caused by end of transmitted byte
      else if(TI0 == 1)
      {
        TI0 = 0;
        tx_complete = 1;
      }
}

//-----------------------------------------------------------------------
void Timer3_ISR(void) interrupt 14
{
      TMR3CN &= ~0x80; // reset timer 3 overflow flag

      /*
        reload = 0xFFFF - ( SYSCLK/12 ) / freq
            = 0xFFFF - ( SYSCLK/12 ) / ( 1 / seconds )
            = 0xFFFF - ( SYSCLK/12 ) / ( 1 / (milliseconds/1000)

          minimum frequency = 28.125 Hz =>  maximum milliseconds = 35.5
      */

      if(timer_remaining_ms > 0)
      {
        // must time in blocks of 35 ms
        if(timer_remaining_ms < 35)
        {
          timer_remaining_ms = 0;
          TMR3 = 0xFFFF - (SYSCLK/1000/12) * timer_remaining_ms;
        }
        else
        {
          timer_remaining_ms -= 35;
          TMR3 = 0xFFFF - (SYSCLK/1000/12) * 35;
        }
      }
      else
      {
        TMR3CN &= ~0x04; // stop timer 3
```

```
      pfnEventCallback();
    }
}

//------------------------------------------------------------------
void send_byte(char byte)
{
    while(tx_complete == 0);

    EA = 0;

    tx_complete = 0;
    SBUF0 = byte;

    EA = 1;
}
```

9

Embedded Microcontroller Based Non-destructive Seafood Inspection System

9.1 Introduction

In this chapter an embedded controller based sensing system for seafood inspection has been described. Interdigital sensors have been used for non-destructive and non-invasive inspection of system properties. There are many applications of interdigital sensors based systems – they are used in bio-medical field to monitor the change in impedance caused by the growth of immobilized bacteria; a micro-sensor based on interdigital electrodes is used to measure the water content in human body as the water content in the skin could be used as an index to confirm the health of human skin. The interdigital sensors can also be used for the estimation of fat content in pork meat and inspect the quality of saxophone reeds. The capacitive sensors can be interfaced with microcontrollers for effective signal processing. Detection of the presence of contaminated acid in seafood has been explained in this chapter.

9.2 Working Principle of Interdigital Sensors

The operating principle of the interdigital sensors depends on the detection of electric field which is altered by the material under test (MUT). The operating principle of an interdigital sensor is the same as that of a parallel plate capacitor; figure 9.1 shows the transformation of parallel plate capacitor to an interdigital sensor. The electric field passes through material under test as it flows from the positive electrode to the negative electrode. Thus, electrode and material geometry as well as material dielectric properties affect the capacitance and conductance between electrodes.

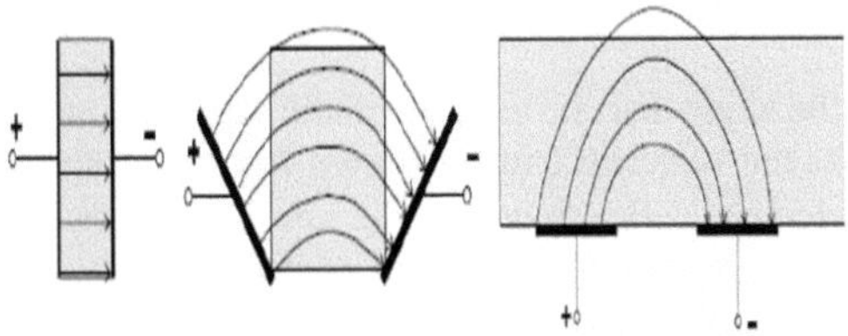

Fig. 9.1 Operating principle of an Interdigital sensor [A]
[A] A.V. Mamishev, K. Sundare-rajan, Y. Du and M. Zahn, "Interdigital Sensors and Transducers", Proceedings of the IEEE, Vol. 92, No. 5, May 2004.

G. Sen Gupta and S.C. Mukhopadhyay: Embedded Microcontroller Interfacing, LNEE 65, pp. 199–213.
springerlink.com

A strong signal can be achieved by repeating the electrode patterns multiple times and this leads to an interdigital structure. One set of electrodes of interdigital sensor is driven by an AC voltage source and the other set of electrodes is connected to ground as shown in figure 9.2. An electric field is formed between the driven and ground electrodes and the penetration of electric field for different wavelengths can be seen from figure 9.3.

The wavelength of an interdigital sensor is the distance between two adjacent electrodes of the same type. In figure 9.3, different penetration depths with respect to different wavelengths of the sensor are shown. When a material is placed in the vicinity of the interdigital sensor, electric fields generated at driving electrodes pass through the most of the material and terminate at sensing electrodes or ground electrodes.

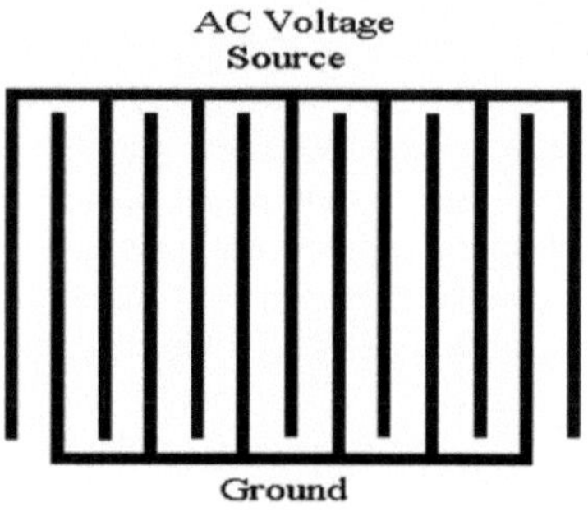

Fig. 9.2 Interdigital sensor structure

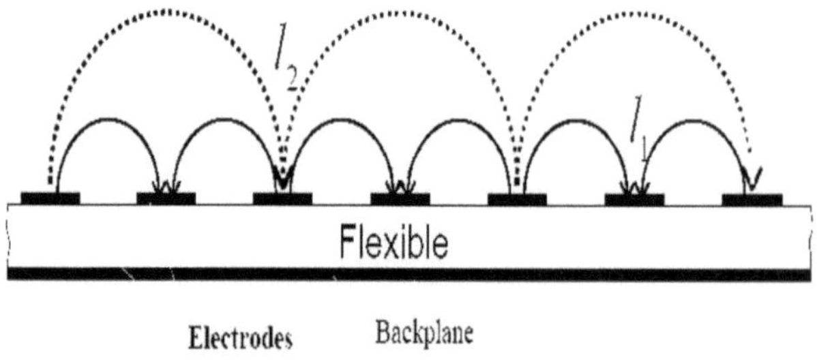

Fig. 9.3 Electric field formed between two electrodes for different wavelengths

The electric field is affected by the dielectric properties of the material under test, hence the dielectric properties of the material can be known from the current or voltage measured at the ground electrode. By changing the area of the sensor, the spacing between the electrodes and also the number of fingers of each electrode, the strength of the output signal can be varied. Figure 9.4 shows the picture of fabricated sensors of different configurations.

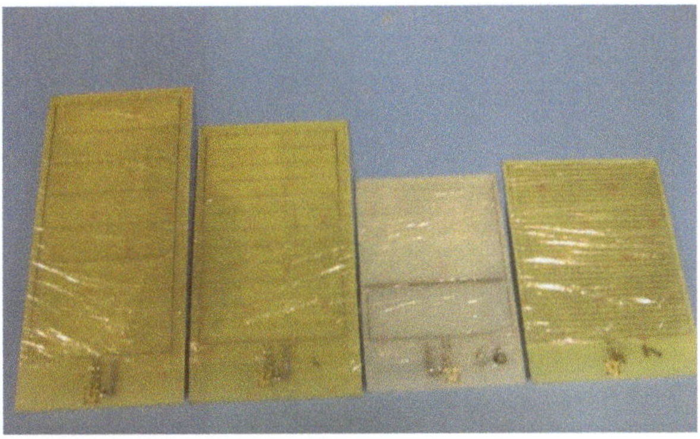

Fig. 9.4 Fabricated interdigital sensors of different configurations

The electrical connection diagram for conducting experiment using the interdigital sensors is shown in figure 9.5. A series resistor is connected with the sensors to measure the current through the sensor.

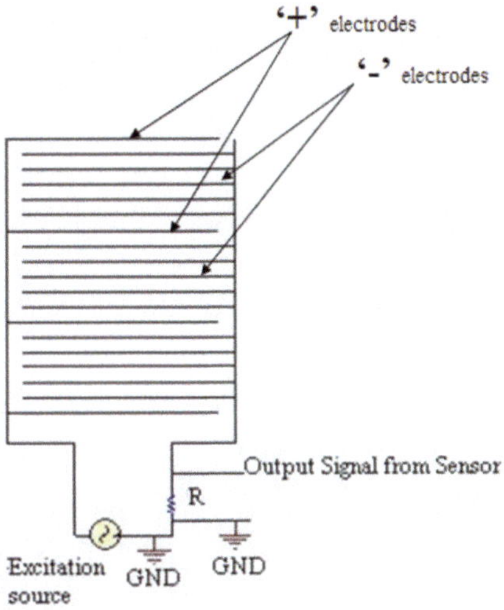

Fig. 9.5 Electrical connection of interdigital sensor for property estimation

The current is measured by measuring the voltage across the resistance, *R*, which is connected in series with the sensor. So we have,

$$V_R = I * R$$

where, V_R is the voltage across R
R is the series resistance
I is the current drawn by the sensor.

Now, $$I = \frac{V}{Z}$$

where, Z is the impedance of the sensor along with R and V is the supply voltage.

$$Z = \sqrt{X^2 + R_{tot}^2} \approx X \quad \text{as } X >> R_{tot};\ R_{tot} = R + \text{resistance of the sensor itself.}$$

$$V_R = \frac{R*V}{X}$$

$$V_R = \omega CRV \qquad \text{as } X = \frac{1}{\omega C}$$

$$= 2\pi f \varepsilon_0 \varepsilon_R AV / d$$

A is the effective area of the sensing element and d is the effective distance between the electrodes.

So, $V_R = K\varepsilon_R f$

where, $K = 2\pi\varepsilon_0 AV / d$, and is constant for a fabricated sensor.

Since $V_R \propto f$ and $V_R \propto \epsilon r$, the voltage across R is proportional to both frequency and relative permittivity.

The output voltages of two different sensors for air, butter, cheese and water at different frequencies are plotted in the figures 9.6 and 9.7 respectively. It can be seen that the sensors have different output values for the same frequencies. However, the nature of the response is similar – the output increases fairly linearly with frequency. The difference in output values for the two sensors can be attributed to the varying pitch lengths and areas. The readings for water were taken by holding the water in a plastic bag and placing it over the sensor which was wrapped in a thin plastic cling wrapper. Butter and cheese blocks were also packed using plastic cling wraps.

It can be observed that the output values for butter and cheese are between those of air and water because their relative permittivity is more than air but less than water. Measures are taken to ensure that there is no moisture content on the sensors or the material under test as this would affect the output of the sensor. To avoid this, sensors are wrapped in thin plastic cling wraps which also helps to eliminate the direct contact of the materials with the sensors.

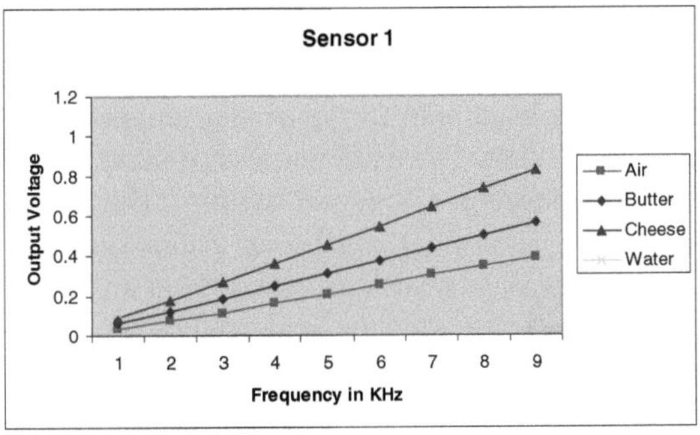

Fig. 9.6 Output voltage of sensor 1 for air, butter, cheese and water

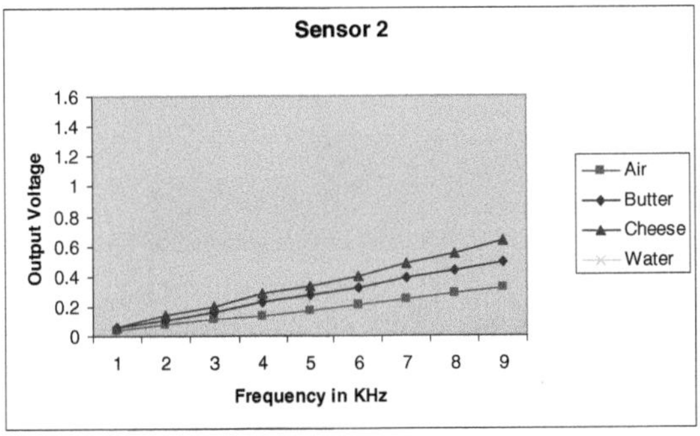

Fig. 9.7 Output voltage of sensor 2 for air, butter, cheese and water

9.3 Sensing System for Seafood Inspection

A low cost sensing system for inspection of raw seafood was developed based on interdigital sensor. The developed low cost sensing system can be used in the fish processing and packaging industry for at a pre-screening stage. The sensing system analyses the samples from the ranch site and then provides a pass or fail analysis. If the results show a certain number of failed analysis (suspicious results), the samples from that ranch site have to be sent to the laboratory for further analysis of contaminated chemicals. Testing for contaminants such as Domoic Acid (DA) in the seafood is an expensive process. The developed sensing system is easy to use for the purpose of sample inspection and can provide fast analysis of DA within shellfish meat for in-situ monitoring. The developed sensing system should be reliable and cost effective.

The developed low cost sensing system consists of a microcontroller, novel interdigital sensor, power supply circuit using 9V battery and signal processing circuit. Subsequent sections will highlight the details of the system called Seafood Inspection Tool (SIT).

9.4 Interfacing to Microcontroller

A SiLab C8051F020 microcontroller was used both for generating the necessary excitation signal as well as for data acquisition. The main purpose of using microcontroller based sensing system is to develop a low-cost system. The microcontroller was programmed to generate a sinusoidal voltage for excitation of the sensor. A sinusoidal waveform of 7.5 V peak-to-peak was generated at an operating frequency of 10 kHz. The stepped sine wave was first generated by the microcontroller and the smoothened sine wave was obtained using smoothening circuit , which was fabricated on a single signal processing board.

The signal coming from the sensor is alternating in nature. The interfacing circuit used for this setup is shown in figure 9.8. The sensor input needs to have an offset since the 12-Bit Analog to Digital Converter (ADC) cannot process values less than zero.

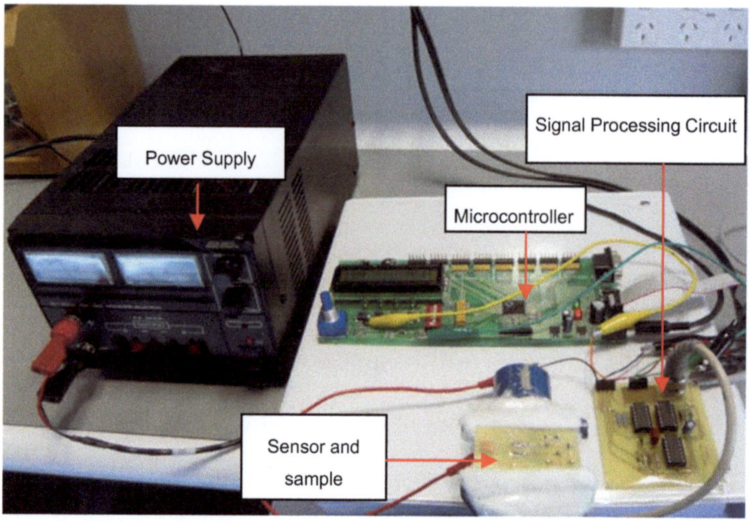

Fig. 9.8 Experimental setup of a low cost sensing system

9.5 Initialization of Important Parts of Microcontroller

The ADC and DAC use a 2.4 V reference by default. The stepped sine wave was first generated by the micro-controller for the excitation voltage. It was generated uisng DAC0 and is connected to a sine-wave smoothening circuit, to obtain the desired excitation signal. AIN0.2 was selected to be the sensor input channel. The initialisation of ADC and DAC are as follows:

```
//---------------------------------------------------------------------
// converter_init
//---------------------------------------------------------------------
//
//Initialise the required ADCs and DACs
//
void converter_init(void)
{
  //--- Set up ADCs & DACs ---
  // Enable the internal bias generator & internal refernece buffer.
  // ==> ADCs & DACs use a 2.4V reference by default
  // Configure the reference for ADC0 to be the VREF0 pin.
  // Turn the temperature sensor off.
  REF0CN = 0x03;

  //*** ADCs ***
  ADC0CF = 0x38;       // SAR0 conversion clock ~2.2 MHz, gain = 1

  AMX0CF = 0x00;       // Configure for 8 single-ended inputs
  AMX0SL = 0x02;       // Select AIN0.2 (sensor input)
  ADC0CN = 0x80;       // Enable ADC0 in continuous Tracking Mode
                       // Conversion initiated on write to AD0BUSY
                       // ADC0 data is right justified
  EIE2 |= 0x02;        // Enable ADC0 interrupt

  //*** DACs ***
  DAC0CN = 0x80;       // Enable DAC0 in on demand mode;
                       // Data is left justified
  DAC1CN = 0x87;       // Enable DAC1 in on demand mode;
                       // Data is left justified

  DAC1H = 0x63;        // Set a reasonable contrast (in case DAC1 is
                       // controlling contrast)
}
```

The sine wave is generated by a sequence of 20 digital samples. The sine wave was programmed using timer 3. Timer 3 is initialised as follows:

```
//---------------------------------------------------------------------
// sine_timer_init
//---------------------------------------------------------------------
//Initialise timer 3 in preparation for using it to generate a sine
// wave
void sine_timer_init(void)
{
  TMR3CN = 0x06;       // Timer 3 uses the system clock; clear its
                       // interrupt flag
  TMR3 = 0xFFFF;       // The timer will overflow straight away
  TMR3RL = 0xFF5F;     // The timer counts from 0x0000 to 0xFFFF
  EIE2 |= 0x01;        // Enable timer 3's interrupt
  EIP2 |= 0x01;        // This interrupt is high priority
  TMR3CN |= 0x04;      // Start the timer
}
```

A brief explanation of the variables used in the program follows- *AIR* is the digital value of the sensor output with air (no sample), *SENSE* is the initial digital value of sensor output with sample, *THRESH* is the threshold value which in this initial design is set to 2048, *NEW_SENSE* is the new digital value of the sensor output and *VIRT_TICKS* is used for the virtual timer. Software initialisation is as follows:

```
//------------------------------------------------------------------
// software_init
//------------------------------------------------------------------
// This function initialises all the global variables
//
void software_init(void)
{
    //--- Sensor Related Variables
    AIR = 0;
    SENSE = 0x0019;
    THRESH = 2048;      // Change the treshold value for DEMO
    NEW_SENSE = 0;
    VIRT_TICKS = 0;
    //--- Sine Wave Generator Related Variables ---
    SINE_SAMPLES = 20;
    SINE_INDEX = 0;
}
```

Hardware initialisation function is used to disable the watch dog timer, to set up the clock, to set up the crossbar and to configure the required LED for the indication of power, pass and fail results. The hardware initialisation is as follows:

```
//------------------------------------------------------------------
// Hardware related initialisation
//
void hardware_init(void)
{
    //--- Disable watch dog timer ---
    WDTCN = 0x07;
    WDTCN = 0xDE;
    WDTCN = 0xAD;

    //--- Set up the clock ---
    OSCXCN = 0x00;              // Don't use any external oscillators
    OSCICN = 0x07;              // Instead use the internal 16MHz clock
    while ((OSCICN & 0x10) == 0);       // Wait for the oscillator to
                                        // stabilize
    //--- Set up the crossbar ---
    XBR2 = 0x40;                        // Crossbar is enabled
    //--- Configure ports 0..3 ---
    //*** OUTPUTS ***
    // Push-pull:       P1.6 (LED)
    //                  P2.0-2 (Indicator LEDs)
    //*** INPUTS ***
    // Open drain:      P3.7 (Start button)

    P0MDOUT = 0x00;
    P1MDOUT = 0x40;             // P1.6 is in push-pull mode
    P2MDOUT = 0xFF;             // P2.0-2 are in push-pull mode
    P3MDOUT = 0x00;             // P3.7 is open drain

    // Write logic 1 to inputs
    P3 = 0x80;
    //--- Configure ports 4..7 ---
    //*** OUTPUTS ***
    // Push-pull:       P7.0 - P7.3 (LCD control)
    // Open drain:      P6 (LCD data)
```

```
    P74OUT = 0x40;
    P2 = 0x01;                    // Turn on the power LED
}
```

9.6 Electronics and Signal Processing Circuit for the Low Cost Sensing System

An efficient data acquisition system is very important in the development of a low cost sensing system. Since the output signal from the sensor is small (in mV) a good circuit has to be designed and developed to minimize the effect of noise. The signal processing circuit consists of a sine-wave smoothening circuit and a signal conditioning circuit. Both circuits were fabricated on a single board as shown in figure 9.9.

The main function of the software is to initialise all the components and then call *introScreen* to let the user run through a complete test or run individual tests via serial control. The main function of the program code is as follows:

Fig. 9.9 Fabricated electronic circuit for signal processing

```
//-----------------------------------------------------------------
// main routine
//-----------------------------------------------------------------
//
// The main function initialises all components ready for testing,
// then calls introScreen to let the user run through a complete test
// or run individual tests via serial control.
//
main()
{
        uint reading = 0;
        bit first = 1;

        EA = 0;                   // Disable all interrupts temporarily
        software_init();
        hardware_init();
        lcd_init();
```

```
        reading_timer_init();
        sine_timer_init();
        converter_init();
        EA = 1;                    // Re-enable all interrupts

        AD0BUSY = 1;               // Start monitoring the sensor

        lcd_reset();
        printf("Press <START>");
        lcd_goto(0x40);
        printf("to calibrate");

        while (P3 | 0x7F != 0x7F)
        {       // Wait for the user
                // wait...
        }

        calibrate();

        lcd_reset();
        printf("Press <START>");
        lcd_goto(0x40);
        printf("to Measure");

        while (P3 | 0x7F != 0x7F)
        {       // Wait for the user
                // wait...
        }

        VIRT_TICKS = 0;
        reading = SENSE;

        while (VIRT_TICKS < 40)
        {       // For ~2 seconds
                if (NEW_SENSE)
                {       // New reading to incorporate in the average
                                NEW_SENSE = 0;
                                reading = (reading + SENSE) >> 1;
                                meter();
                }
        }
        reading = SENSE - AIR;
        lcd_reset();
        lcd_goto(0x40);
        printf("Press <RESET> ", reading);
        lcd_goto(0x0C);
        if (reading > THRESH)
        {
        // failed
                printf("FAIL");
                P2 |= 0x14;             // Turn on the fail LED
        }
        else
        {
                // passed
                printf("PASS");
                P2 |= 0x40;             // Turn on the pass LED
        }
        while (1)
        {
                                        // stop the program here
        }
        return(0);
}
```

9.7 Smooth Sine Wave Generation

The smoothened sine wave was obtained using a smoothening circuit. The circuit diagram of the sine-wave smoothening circuit is shown in figure 9.10. The step sine wave generated by the microcontroller has a very high frequency component; therefore a low pass filter of 6 kHz is needed to make the sine wave smooth. A decoupling capacitor was used to reduce noise. A pull down resistor of 1 MΩ was used to bring the sine wave at symmetry with respect to ground. The 100 kΩ resistor is used to minimize the current input to the non-inverting op-amp. The sine-wave generated by the micro-controller before and after the smoothening circuit is shown in figure 9.11. The generated sine wave was used as an excitation signal for the developed novel interdigital sensor.

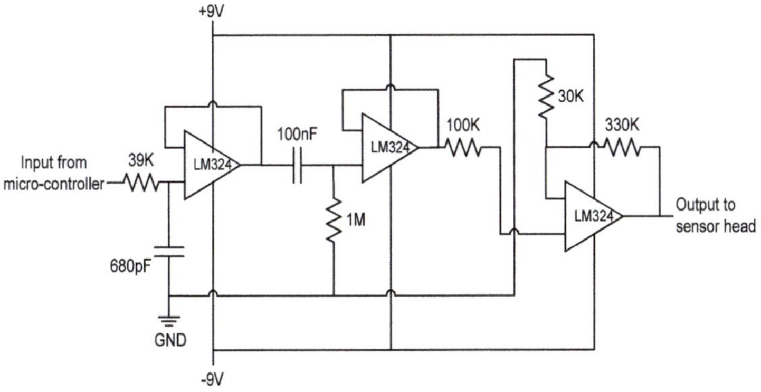

Fig. 9.10 Sine-wave smoothening circuit diagram

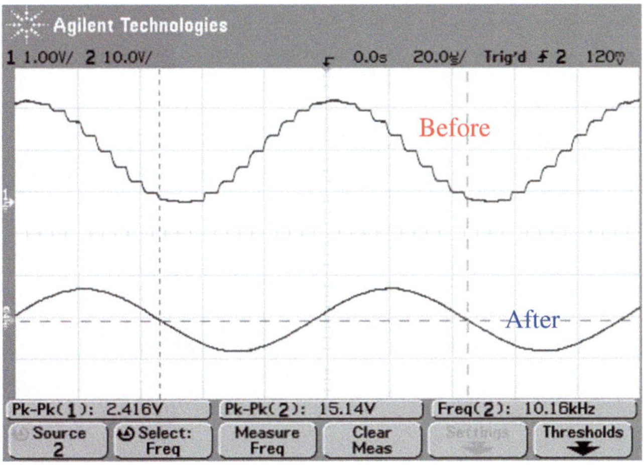

Fig. 9.11 Sine-wave generated by the micro-controller before and after the smoothening circuit

9.8 Signal Rectification and Amplification

The second part of the signal processing circuit is the signal conditioning circuit which was built using several operational amplifiers. It is used to interface the sensor signal to the microcontroller as shown in figure 9.12. It consists of a full wave rectification circuit and an amplification circuit.

The operation of the circuit is as follows: the output voltage from the sensor is fed to the first op-amp (unity gain amplifier) to generate a full sine wave signal (positive and negative). The supply for this op-amp is bipolar whereas all the other op-amps have uni-polar supply. If the input voltage form the sensor, V_{SEN}, is greater than 0V, then the output from Op-amp 2 equals V_{SEN} but only the positive half of the input signal is available at the output. Op-amp 3 operates as a substracter, delivering an output voltage which equals two times the output of op-amp 2 minus the output of op-amp 1. In effect the output of op-amp 3 is the rectified version of the input signal. The rectified voltage will pass through op-amp 4 with a gain of 9.2. The amplified signal will pass through a low pass filter with a cut off frequency of 13 Hz. The dc signal from op-amp 5 is connected to a digital input of the microcontroller for the necessary conversion into digital value.

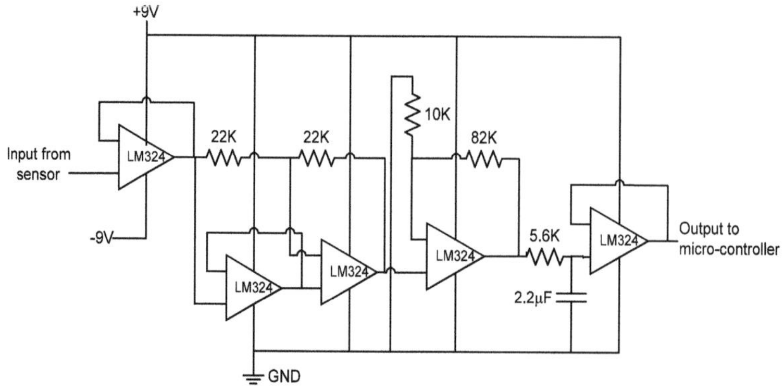

Fig. 9.12 Signal conditioning circuit

9.9 Calibration, Sensitivity Threshold and Signal Definitions

The sensing system needs to be calibrated before it can start taking measurements and provide analysis of the samples under test. The calibration is based on the digital value of the sensor output at air, ADC_{AIR} (sensor without sample). The microcontroller reads this data first and uses it for calibration. The sample with known thickness is placed on the sensor for measurement. The micro-controller then records the sensor reading in terms of digital value of that sample, ADC_{OUT}. The sensitivity of the sensor is given by-

$$Sensitivity = \frac{(ADC_{OUT} - ADC_{AIR})}{ADC_{AIR}} * 100$$

where

ADC_{AIR} = Digital value of the sensor output at air

ADC_{OUT} = Digital value of the sensor output with sample

The current sensitivity value is compared to the required threshold sensitivity to get a pass or fail analysis of the particular sample.

```
//------------------------------------------------------------------
// calibrate
//------------------------------------------------------------------
//
// Calibrate the sensor by sensing the air for ~2 seconds (to get a
// good average reading)
//
void calibrate(void)
{
        uchar count = 0;
        bit flash = 0;

        lcd_reset();
        printf("Prepare to");
        lcd_goto(0x40);
        printf("calibrate sensor");
        huge_delay(50);

        while (P3 | 0x7F != 0x7F)
        {       // While the user hasn't responded...// Flash text
                if  (!flash)
                {       // If flash is off...
                        lcd_goto(0x00);
                        printf("Press the button");
                        lcd_goto(0x40);
                        printf("   when ready   ");
                }
                else
                {
                        lcd_reset();
                }

                large_delay(122);// Check the button every 10ms or so

                if (count == 25)
                {       // Flash on for 35ms
                        count = 0;
                        flash = ~flash;
                }
                count++;
        }
```

```
        lcd_reset();
        printf("Calibrating...");
        AIR = SENSE;
        T4 = 0;
        VIRT_TICKS = 0;

        while (VIRT_TICKS < 40)
        {       // For ~2 seconds
                if (NEW_SENSE)
                {       // New reading to incorporate in the average
                        NEW_SENSE = 0;
                        AIR = (AIR + SENSE) >> 1;
                        meter();
                }
        }
}
```

9.10 Prototype of Seafood Inspection Tool (SIT)

The first prototype of seafood inspection tool (SIT) was developed to detect domoic acid (DA) in mussels. SIT consists of a ± 9V power supply, a novel planar interdigital sensor, a SiLab C8051F020 microcontroller, a signal processing circuit and an expansion board (for display). A user friendly software was developed to make it easy to use. It can be used by anyone, especially by fisherman, for pre-screening process at the ranch site. The first prototype of SIT is shown in figure 9.13.

Fig. 9.13 Seafood inspection tool (1st prototype)

9.11 Conclusion

A smart sensing system for food monitoring was developed to assess seafood contaminated with marine bio-toxins. A low cost sensing system, using a microcontroller and a few fabricated circuit boards, has been developed. This chapter has discussed the development of the various components of the sensing system. The first prototype of SIT was introduced to help the fishermen to conduct the pre-screening process for the detection of domoic acid. If the results from the samples are suspicious, the whole batch should be isolated and detailed laboratory analysis, using expensive equipments, should be conducted.

MIX
Papier aus verantwortungsvollen Quellen
Paper from responsible sources
FSC® C105338

If you have any concerns about our products,
you can contact us on
ProductSafety@springernature.com

In case Publisher is established outside the EU,
the EU authorized representative is:
Springer Nature Customer Service Center GmbH
Europaplatz 3, 69115 Heidelberg, Germany

Printed by Libri Plureos GmbH
in Hamburg, Germany